BACKYARD CHICKENS

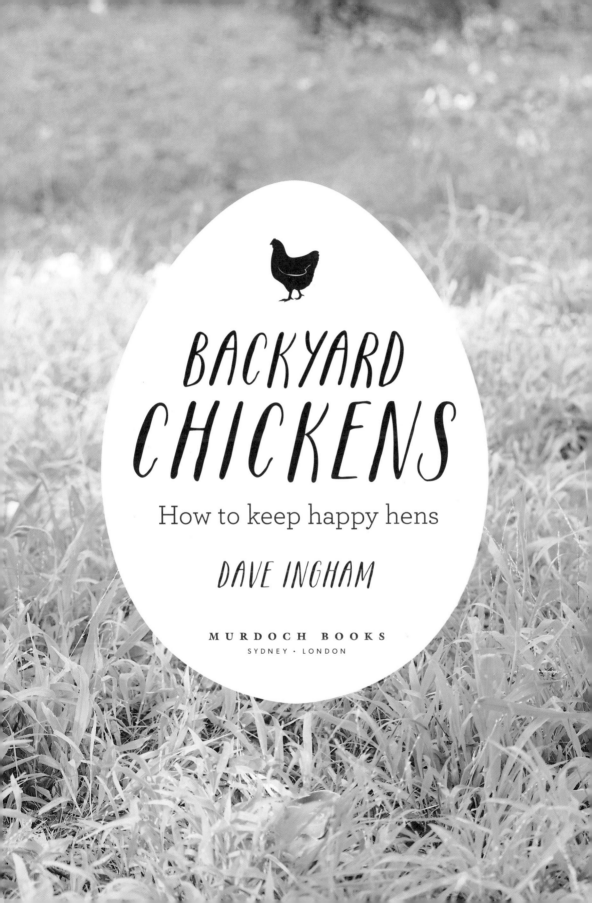

BACKYARD CHICKENS

How to keep happy hens

DAVE INGHAM

MURDOCH BOOKS
SYDNEY · LONDON

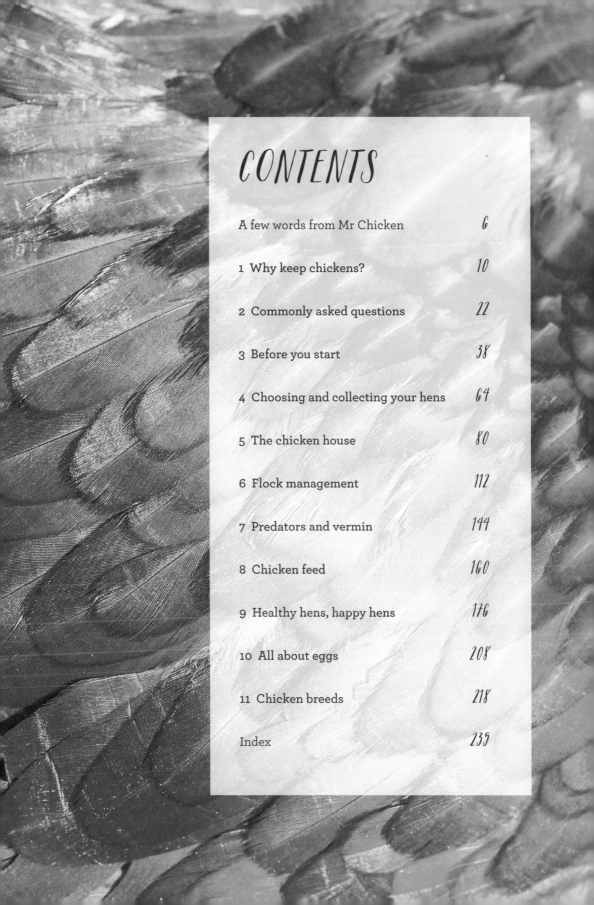

CONTENTS

A FEW WORDS FROM
MR CHICKEN

I have always loved birds. I love watching them move. However, it wasn't until university that I first kept birds as pets or had any relationship with chickens other than my surname (synonymous with poultry in Australia). Although I am not related in any way to those other Inghams, should they choose to shower me with largesse, I'm sure I could find a familial link, however tenuous.

My grandparents kept chickens, but I only saw them three times because they lived on the other side of the world. I recall my grandmother bartering her fresh eggs for dry cleaning in the local town, and especially remember the first time I held a warm, just-laid egg, but not much else.

At university I lived in a ramshackle share-house on a huge block that was rented to students while it awaited demolition, subdivision and redevelopment. I inherited being a foodie from Mum and expanded my culinary options by growing as much food as possible in the vast garden available to me. Chickens were a natural addition. This all predated the Internet, of course, so I turned to the

local library for books on keeping farm animals and to the newspaper classifieds for the hens themselves. My first coop was made of a wooden machinery packing case and some fencing mesh scavenged from a neighbourhood clean-up. It lasted for years and was exactly the right price for an impecunious university student.

A decade or so later I was asked to speak at a sustainable living project about modifications I'd made to my house. There I was, waxing lyrical about the sustainability of chickens as pets, when, out of the blue, I offered my spare coop and a couple of hens on loan to the other participants. Someone coined the term 'Dave's rent-a-chook' and when the project finished I was photographed holding two hens. A piece in the local paper turned into an article in a national newspaper and two radio interviews and I thought, with this much free publicity, there's got to be a buck in it somewhere. A casual conversation with my friend Pete, some scribbling on a sheet of butcher's paper on his kitchen table, and Rentachook was born.

In the 15 years since, Rentachook has had its ups and downs but most of all, it's been a hoot. As well as becoming the go-to spokesperson for all-things-chicken, I have met amazing people who overcame their initial trepidation and came to share my passion for keeping hens.

Selling chickens, you can't take yourself too seriously. Beware: hens are sometimes known to spontaneously grow sharp little teeth that should be filed down before the birds become bloodthirsty. For many years Rentachook sold tooth files for the management of this poorly researched and little understood condition. The tooth files had quite special attributes, being both invisible and weightless. A bargain at just $10 each and all proceeds given to charity.

My favourite tooth file sale was to a guy I'll call Damien (not his real name). Damien ordered his coop and hens online, including extra feed and a tooth file. How I loved checking orders and finding one with a tooth file request.

Everything was delivered and all went well. However, a week later I got an email from Damien. He was thrilled with his new chickens, but couldn't find his tooth file. I replied with an apology and said I would send him one. I posted an empty padded envelope. Another polite email from Damien arrived later that week, advising that the envelope had arrived, but was empty. My reply included a link to the description on the website, clearly advising that the tooth files were invisible and weightless.

Another polite enquiry from Damien – he didn't understand. I was on the point of sending him another empty envelope when I thought, I have to let him go. So I sent the reply, 'Have you ever heard the expression "rare as hens' teeth?"'

Damien: 'Oh.'

Five minutes later, Damien again: 'Got me.'

No sense of humour? Chickens aren't for you.

Writing a book was a bigger undertaking than I first thought. Many thanks to Jane, my publisher, who persevered until I committed to it. Many thanks also to my mate Chris Allen, the author of the *Intrepid* books, who promised to throttle me if I didn't see it through. He's an ex-paratrooper, so that provided motivation.

The most thanks of all to my family: to my three sons, Hamish, Alex and Ted, born in consecutive years, and my wife, Catriona, who makes me who I am and shares my penchant for not taking me seriously.

1

WHY KEEP CHICKENS?

I t might read as henvangelical, but I love keeping chickens
and here are just a few of the reasons why I think everyone
else should do so too.

Fresh free-range eggs:
the best you've ever tasted

For my mind the showstopper of the chicken-keeping experience
is the flavour of the eggs. Commercially produced eggs,
especially cage eggs, come from hens that are, by necessity, fed
a homogenous diet. This results in consistent eggs, but they
lack the richness and depth of flavour that you get in eggs from
backyard hens.

Backyard hens are free to eat whatever they fancy – and this is
a wider diet than you might expect, even if they're not fed kitchen
scraps. Hens are omnivorous, meaning they eat anything. Meat,
vegies, weeds, grubs and bugs all go together to give you a really
rich egg that is properly tasty to eat as well as looking tasty.

I once made two versions of scrambled eggs, using identical
quantities and cooking methods, and served them up to friends.
The only difference between the two was that one pan used

Opposite: Collecting
eggs is a highlight of
the chicken-keeping
experience and it
fascinates kids.

home-grown 'Dave eggs' and the other, commercial eggs. Immediately, you could see that one dish was richer in colour and just looked far yummier. The proof was in the eating and, while the Dave eggs were wolfed down, most of the other dish was left over. When I explained what I'd done, everyone agreed that there was just something special about the home-grown eggs.

Also, just for fun, I have been known to take my eggs to cafés and ask that they be made into eggs benedict (my favourite breakfast). I've never had my request turned down, but I have had the chef come out of the kitchen (chef's hat, apron and all) and ask where I got them, because 'I haven't seen eggs like those in years'.

Eggs, like all foods, degrade with time and fresh is best. As eggs age, the consistency of the white (albumen) changes from firm to runny. This means that when you poach a really fresh egg it holds together well. There is nothing quite like running up to the kitchen to poach an egg that is still warm from the hen.

Ethical farming and knowing where your food comes from

As concern increases about additives such as nano-particles, preservatives, flavourings and sugars in our food, so does the importance of knowing the origins of what we eat.

I also advocate that an animal shouldn't have to live a life of suffering so that the price of the food it provides is reduced. I personally believe that the home-grown egg tastes better because the chook is happier, living a proper life for a chicken, free of stress and confinement — I have no concrete evidence to back this belief, but it does ring true, doesn't it?

I recall a radio interview when I was asked if I thought the higher price of certified organic eggs was warranted – whether organic free-range eggs cost too much? I countered by saying that the price of cage eggs is too low and the difference in the price between cage eggs and free-range eggs is paid by the chicken, not the purchaser.

If reading this book doesn't convince you to keep chickens yourself, I will consider the book a success if it convinces you never to buy cage eggs again.

Opposite, clockwise from top left: Speckled Sussex; eggs from ISA Browns; Silkies make great pets but don't expect many eggs; nothing says 'nest' quite like golden straw.

Backyard Chickens

A revenue-positive pet

Once you have covered the initial cost of setting yourself up with chickens, the cost of feeding and maintaining them is comfortably less than that of buying eggs, especially if you usually buy proper free-range/organic eggs. Unless you splurge on some extravagant *'palais de poulet'*, it should take between 12 and 18 months to pay off your initial costs. After that, the cost of feeding and maintaining your flock will be amply compensated by the savings you make in not buying eggs. More on this later.

Chickens are environmentally sustainable and greener than Kermit's bum!

Being omnivorous, chickens are happy to eat all sorts of food waste that would otherwise end up in landfill. Even experienced composters struggle to compost meat, pasta, bread and sauces. Unless you have a particularly greedy (and usually rotund) dog, such foods can't be recycled at home and end up in the bin.

To satisfy curiosity, I once spent three months weighing the food scraps I fed to the hens. Granted, I have a fair-sized flock (usually between six and 12 birds), but I was giving them around 10–15 kg (20–35 lb) of food waste per week. That's a lot of waste diverted from landfill, from my suburban house alone. I don't ever suffer the enviro-guilt of food wastage – because my excess food isn't wasted, it's just eaten by a chicken instead of a person.

I'm sure you've heard of the concept of food miles. Well, think of the path an egg takes from a farm to your fridge. The egg laid at a farm somewhere in a rural area is then transported by truck to a facility for grading, sorting and packaging in a carton. The cartons are then transported by truck to a supermarket distribution centre, usually on the outskirts of a major city. At the distribution centre the carton is put onto a pallet to be distributed, again by truck, to your local supermarket. At the supermarket you add it to your shopping, which, if you are like most people, you transport home in your car. I make that at least three journeys by truck and one by car and, given that most commercial egg producers benefit from efficiencies of scale by using big regional facilities, the distances travelled can be vast: many hundreds of kilometres.

Now think of the food miles for your backyard egg. No, scratch that and think in food metres... In my case, about 50 metres.

16

Environmentally sustainable pet = low food miles

Commercial egg journey

Home-grown egg journey

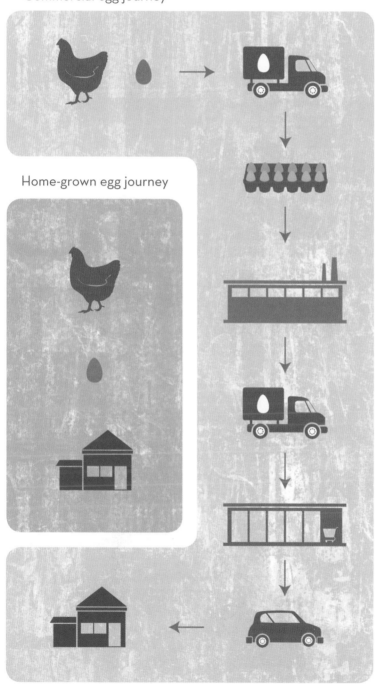

Cheap garden maintenance

Chickens will remove weeds from your garden and can turn your compost heap over for you. I have seen hens fight over a cockroach, beetle, slater or worm. They are obviously quite happy eating leafy grasses, but they will also voraciously attack your garden weeds.

Chickens can be used actively or passively for weed control and at the end of a growing season they'll turn over your vegetable bed in preparation for the next season. And, if you live in an area where there are problems with ground-dwelling insects such as spiders or ticks, chickens can be used as an alternative to spraying because they will eat any insect they can reach.

A low-maintenance pet

If you have kids pestering you for a cat or dog, give thought to the poultry alternative. Many people hesitate to get pets for fear of the time involved in training and exercising them, and the constraints on family holidays.

Not so with chickens: no training required, as much or as little interaction as you like, and they're easy to care for when you're away from home.

With dogs you can have problems of boredom and lack of exercise – hens are different. Despite vociferous protestations from a vocal few to the contrary, there really just isn't enough brain inside a hen to get bored. With a small area to scratch around in looking for insects or seeds, chickens will get plenty of entertainment and all the exercise they require.

With cats comes the ever present risk to native wildlife posed by keeping a top-order predator in confined conditions. (Cat-owning friends have advised me that this risk is reduced by attaching a bell; however, if the bell is not the size of a cow bell and cast from bronze, I'd say they are overstating its efficacy.) Chickens don't hunt native animals (insects not included).

My experience is that chickens are quite happy doing their own thing if you need to leave them to their own devices. Just cover the basics (food, water, shelter and protection from predators) and hens look after themselves.

If you're spending more than 5–10 minutes a day looking after your chickens, you are overdoing it. I don't mean to discourage

Opposite: A Barnevelder and an ISA Brown doing some garden weeding. Both varieties have lovely calm temperaments.

you from enjoying or interacting with your chickens, I'm just saying that the mandatory care of them is minimal.

Let them out in the morning, top up the feed and water, collect the eggs (3–5 minutes tops) and then lock them up at night (2 minutes). Too easy. If you go away for a long weekend they can be left cooped up (assuming you have a suitable sized coop) and if you go away for longer, deputise a neighbour and bribe them with eggs. How hard can it be?

Kids love them, too. Girls will invite them to tea parties and into a cubby house. And boys? Well, the problem I have with my three sons is stopping them playing with their chicken friends on the trampoline or sliding them down the slippery dip.

Note: No hens were harmed in the writing of that paragraph.

Opposite: If your children are pestering for pets, hens are a great option.

2

COMMONLY ASKED
QUESTIONS

Can I keep chickens in my backyard?

Allow me to dispel the rumour that chickens are farm animals and only suited to a rural setting – that is just not the case. In many ways chickens are better suited to urban and suburban homes than dogs or cats. With the possible exception of apartment blocks, I have seen hens kept in every variation of dwelling with great success. (I've even seen people in apartment blocks keep hens in a communal area with consent of their neighbours.) The art lies in ensuring the size of the flock is appropriate for the location available. Given that two hens can be kept happily and indefinitely in a run the size of the shadow under a large car, most households can keep hens – go on, give it a go.

A lot of people I have spoken to over the years have had a preconception that chickens will smell, attract snakes or rats, or destroy their garden. While it is true that if you fail to manage any activity properly you will have problems, there is no reason that backyard chicken-keeping should give you grief. Let me quickly shoot down a few common misconceptions...

Opposite: This super-cool chicken tunnel provides access for the hens and also discourages wild birds from entering the coop and stealing food.

Chickens are smelly

Not so. However, if you have too many hens in too small an area and fail to clean it out occasionally, yes it can get smelly. But that is why you are reading this book, right? To learn how to avoid such pitfalls.

Chickens attract snakes

I don't know where this idea comes from. In more than 20 years of keeping chickens I have never seen a snake. Moreover, in more than 14 years of selling chickens, I have never had a customer complain about snakes. Granted, my experience is limited to Australia, but I'd say I have the wide brown land pretty well covered and we're famed for our snakes.

Chickens attract rats

Nope, not true either. Spilt food, especially if left overnight, will attract vermin. Manage the food well and you won't have vermin problems. See Chapter 7 for an elaboration.

Chickens will destroy my garden

Yes, hands-up, you've got me there. Actually, I might be going a bit overboard on the mea culpa here. Hens will go through your vegie patch like a rotary hoe, if you let them. However, there is a simple solution: if you are worried about your chickens harming your garden, either fence in the garden or fence in the birds – it is called a hen run. You can let them out at leisure, but if they spend most of their time in a hen run, the impact on your garden will be negligible.

And, yes, keeping hens in a suburban setting *is* legal

There are occasional restrictions, but in all the localities I have looked into chicken-keeping is permitted and, in many cases, encouraged by local authorities. The requirements revolve around ensuring the flock is well kept (free from odour, vermin and roosters who make noise) and ensuring that you don't go overboard with your new passion and turn a household flock into a commercial business.

Opposite: A beautiful silver-laced Wyandotte finds a backyard vantage point from which to survey her domain.

Is my space going to work?

Over the years I have had many prospective customers request that I provide a *pre-chicken suitability inspection* of their property and for a while I was willing to perform this service. However, it soon became apparent that with very few exceptions, every single urban and suburban property I visited or delivered chickens to was suitable, or could be made so with minor accommodations. Here are a few examples:

An inner-city terraced house

Think: close in to the city centre, a row of very small houses, joined together, with pocket-sized gardens which, because they are so small, often tend to be fully landscaped.

This house had a narrow passage (maybe 1.5 metres/5 feet wide) running between the boundary fence and back of the house. The area was unused and overgrown. By fencing the end of the passage, the chickens had 8 square metres (85 square feet) of space without encroaching on the landscaped part of the garden. This space was only suitable for two hens, but they thrived.

An inner-city apartment block

Close to the city centre on a small block of land, this place was an old walk-up block of four apartments in one building with a bit of grass that was unloved and mostly in shadow. One couple started it off by convincing their neighbours into a six-week trial of two hens and a small coop. Despite initial reservations by some, by the end of six weeks the coop had been upgraded to a bigger one and two more hens added (another family bought in). Within a year, there were eight hens, two for each household, and a real communal approach to managing the chickens. The mangy grass had been replaced by a hen run that was covered in mulch and looked great, and the hens were as happy as clams.

A suburban house with a very steep, overgrown and rocky garden

This is one of my fondest memories of supplying chickens. I had to spend more time than I ever had before overcoming the reservations of the lady who came to me – she was both very reluctant and very keen at the same time. I finally convinced her to give it a go and trial a couple of hens. She became besotted with the chickens and 18 months later had a flock of 12 birds, got a carpenter in and built a true *palais de poulet*.

If you can answer yes to the key questions on page 36 and get to the end of the quiz, I reckon chickens will work at your place. If you are still sitting on the fence, please bear in mind that the set-up costs are not huge and that, unlike a cat or a dog, chickens need not be for life. You can sell them or give them away without suffering paroxysms of guilt.

What is it likely to cost?

The costs associated with keeping chickens are probably easiest to understand if broken up into:

- Set-up costs
- Ongoing costs and
- Problem-solving costs

Set-up costs

Your major set-up cost is the coop. Whether you call it a henhouse, coop, chicken-tractor, chook Hilton, roost, or *palais de poulet*, the chickens are going to need shelter and protection from predators at night, especially the altogether too talented Mr Fox. There is a lot more detail about coop options in Chapter 5, but, assuming you are not a handyperson (and let's be frank, who is these days?) you will be buying one.

In Australia, where I'm writing, prices for coops range from a couple of hundred dollars for an imported flat-pack timber job from an online marketplace, to A$400–800 for a quality, locally made, portable coop or 'chook-tractor'. An alternative to this is a galvanised pressed-metal aviary, ranging from A$250 to A$600 for a fair-sized one to get you started. So, we're not talking megabucks. Prices in the US and UK are comparable.

The next expense is the hens themselves. Understandably, hens are cheaper if you buy them in the country rather than from a city pet shop or stockfeed supplier. For a common crossbreed laying hen, at point of lay, you can expect to pay A$30–35 in the city and as little as A$15 in the country. Of course, you can pay silly money for show-quality, purebred hens but it is my recommendation for first-timers to start with a crossbred hen.

Other set-up costs can include feeders, waterers and storage bins for feed and straw. However, these cost in the few tens of dollars and the Parsimonious Poultry Person could make these themselves by reusing plastic milk bottles and the like.

If you want a hen run, a simple one can be made very cheaply at home from chicken wire and tomato stakes.

So, add it all up and you can easily have two hens and all their mandatory accoutrements for less than A\$400, or the same thing but with a good-quality coop for under A\$600. Let's split the difference and say, set up for A\$500.

Ongoing costs

Feed is your regular ongoing cost for chickens; expect to spend in the order of A\$10 a month on it for two hens. Really, that's chicken feed! (Sorry, I couldn't help myself.) Depending on where you live, that's about the cost of two to three large takeaway coffees. Add another A\$5 a month for straw and quarterly worming... Assuming you keep an ongoing flock of around four hens, and rounding it up a bit, let's say A\$30 per month upkeep.

Problem-solving costs

The reason you are reading this book is so that you can pick up some canny tricks and techniques to solve problems inexpensively. In truth, with good management, chicken problems that require other than negligible expenditure are rare. Hens don't require regular vet check-ups and by-and-large vet visits, when required, are not prohibitively expensive.

That being said, I had a regular customer many years ago who told me about her extraordinary veterinary experience. Diamond and Pearl (*not their real names*) had gone off the lay and were not their usual selves. She took them to a particular vet (who shall remain nameless to protect the guilty). Tests were performed to diagnose their ailment. Suffice to say, they received excellent care and were returned in fine health, but not before a bill of nearly A\$3000 had been accrued! 'Three grand? Are you kidding me?' I had to sit down when I was told this story.

Bear in mind that this was for two ISA Brown hens that were identical in almost every way to the hens I could have sold her (to replace Diamond and Pearl) for A\$30 each.

The moral of this story is: keep some perspective, people.

Opposite: A bright-eyed ISA Brown. Overleaf: My three young cockerels holding, from left, a silver-laced Wyandotte, an Australorp and a gold-laced Wyandotte.

While your hens become pets and quasi family members, they are still farm animals at heart and can be easily replaced. Although chickens can get diseases that can spread through a whole flock, in a backyard you are not going to have enough hens to warrant spending your kids' inheritance on vet's bills. Now, if it were a A$20,000 macaw, that would be a different story, but with a A$30 sick hen, culling and replacing is a reasonable approach that shouldn't leave you wallowing in guilt.

Payback time

Let's assume your family loves its eggs and you normally buy a couple of dozen a week. Therefore, you decide to keep four hens.

Let's also assume that, like me, you abhor the unconscionable conditions in which cage eggs are produced. And, also like me, you believe in the benefits of organic farming which, for hens, means free-range and significantly lower stocking rates per acre. This means that you are currently paying in the order of A$8 per dozen for your eggs, a total of A$832 per year! Quite a lot to shell out (*pun intended*) just on eggs when you look at that number, isn't it?

Now for the show stopper: A$30 per month upkeep means A$360 per year outlay for your eggs and a saving of a whopping A$472 per year on egg costs. Enough to pay off your set-up costs in the first year. After that, it is happy days.

Not only do you get the best eggs you've ever tasted but keeping hens saves you flipping-great wads of cash as well.

Wrap it up, I'll take it, you say: chickens are for me!

What if it doesn't work out?

The original idea behind Rentachook was to encourage people to dip their toes into keeping hens without needing to commit to having them permanently. You get six weeks to see if it's going to work and, if not, you can hand back the birds and all their paraphernalia.

Obviously, if you are uncertain, this model or something like it is a great way to commence your journey into backyard poultry farming, as you have a built-in exit clause. (After reading this book you will, of course, be properly informed and therefore won't require an exit clause.)

However there are a number of circumstances in which keeping hens is no longer suitable. These can include relocating for work, major renovations that require moving out, or extensive landscaping. Whatever the reason, you have a few live animals that need a new home.

While I have known people who claim to have a special relationship with their chickens, my experience is that, if provided with their basic needs, hens will live just as happily in someone else's garden as in yours and will readily accept relocation. Unlike a dog or cat, your hens will not pine for you and don't expect a reunion like John and Ace had with Christian the lion should you go and visit them. This should hopefully assuage any guilt you might have a tendency to feel about foisting your pets on friends or strangers.

The best option for rehoming your hens is adding them to the flock of a friend, family member, someone you know through school sport or whatever. No harm, no guilt and you might even be able to cadge the odd egg from them.

If that fails, advertise the hens and coop as a package, perhaps 'free to a good home'. After all, you aren't going to need the coop without the hens are you?

You will find it a lot easier to find a home for younger hens that are still laying than for older non-productive hens. In truth, nobody who keeps chickens for eggs wants additional hens that are nearing the end of their laying or past laying. You may have to bribe your recipient into taking your older hens. I'm told a case of beer is good currency in some quarters.

As a last resort, if your hens are old and your coop not worth relocating, the hens can be culled. As 'culled' actually means 'killed', this should only be considered if there is no other option.

The 'Can-I-Keep-Chickens?' Quiz

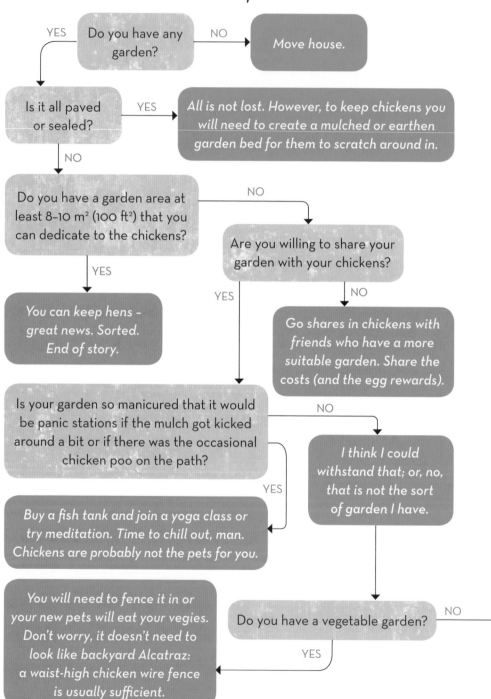

YES

Do you have any garden?

NO

Move house.

Is it all paved or sealed?

YES

All is not lost. However, to keep chickens you will need to create a mulched or earthen garden bed for them to scratch around in.

NO

Do you have a garden area at least 8–10 m² (100 ft²) that you can dedicate to the chickens?

NO

Are you willing to share your garden with your chickens?

YES

You can keep hens – great news. Sorted. End of story.

NO

Go shares in chickens with friends who have a more suitable garden. Share the costs (and the egg rewards).

YES

Is your garden so manicured that it would be panic stations if the mulch got kicked around a bit or if there was the occasional chicken poo on the path?

NO

I think I could withstand that; or, no, that is not the sort of garden I have.

YES

Buy a fish tank and join a yoga class or try meditation. Time to chill out, man. Chickens are probably not the pets for you.

You will need to fence it in or your new pets will eat your vegies. Don't worry, it doesn't need to look like backyard Alcatraz: a waist-high chicken wire fence is usually sufficient.

Do you have a vegetable garden?

NO

YES

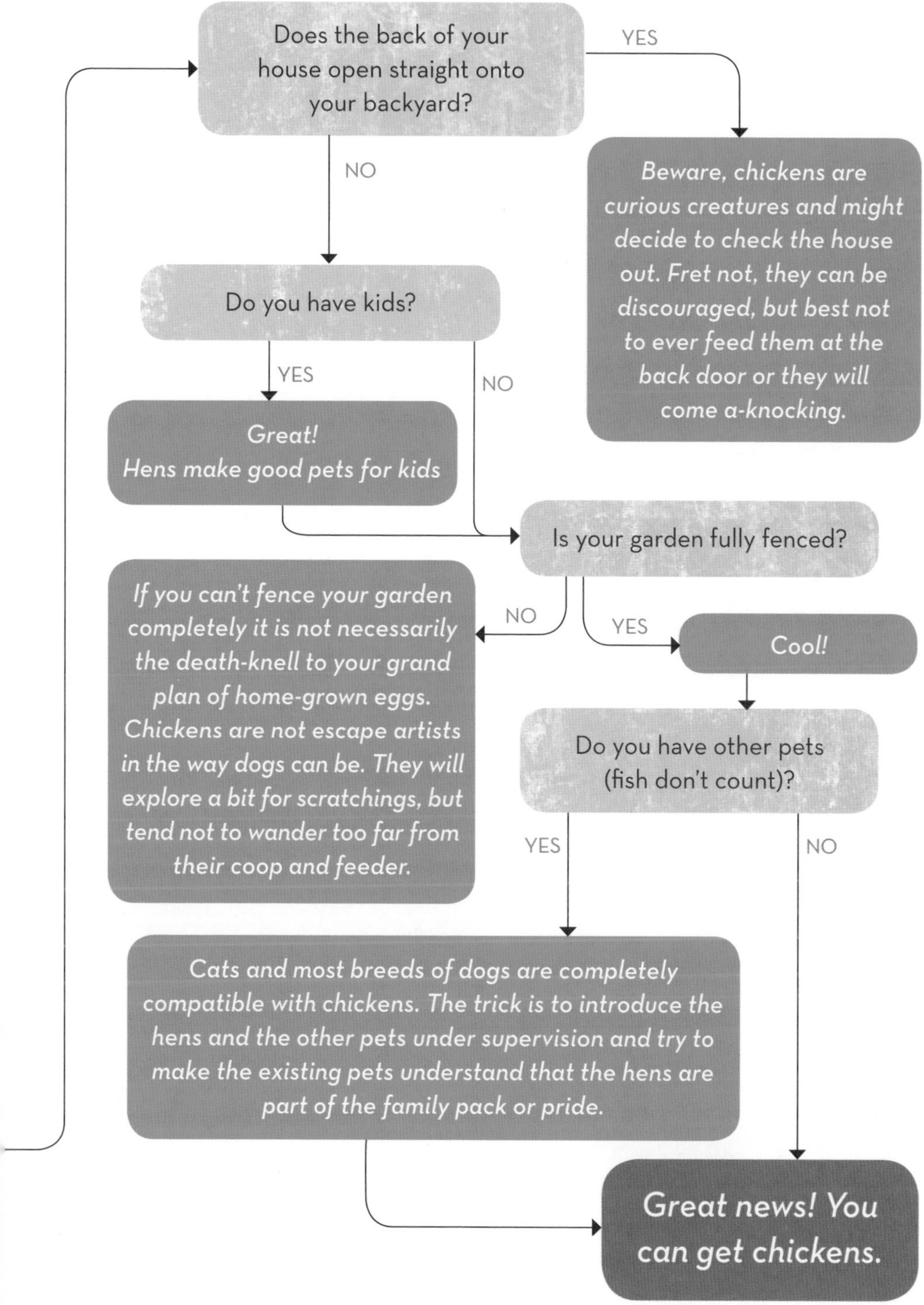

Does the back of your house open straight onto your backyard?

YES

Beware, chickens are curious creatures and might decide to check the house out. Fret not, they can be discouraged, but best not to ever feed them at the back door or they will come a-knocking.

NO

Do you have kids?

YES

Great!
Hens make good pets for kids

NO

Is your garden fully fenced?

NO

If you can't fence your garden completely it is not necessarily the death-knell to your grand plan of home-grown eggs. Chickens are not escape artists in the way dogs can be. They will explore a bit for scratchings, but tend not to wander too far from their coop and feeder.

YES

Cool!

Do you have other pets (fish don't count)?

YES

NO

Cats and most breeds of dogs are completely compatible with chickens. The trick is to introduce the hens and the other pets under supervision and try to make the existing pets understand that the hens are part of the family pack or pride.

Great news! You can get chickens.

3

BEFORE YOU START

Much like returning home from hospital with babe-in-arms, there are some things you really need to sort out and/or acquire before you set off to collect your hens.

There are some businesses these days (and I have to disclose an interest here) who will set you up with the full kit-and-kaboodle – bringing around to your house everything you need for keeping a small flock of hens. However, it is also fine to do this yourself. And a little pre-planning will increase the likelihood of a successful poultry venture.

What to buy? Chicks or chooks? Purebreeds or crossbreeds?

For the first-time chicken-keeper, I always recommend starting with a Volvo. What? A Volvo? What the feather does a Volvo have to do with chickens? Read on.

When you teach someone to drive, you don't put them behind the wheel of a Ferrari, do you? Well, not if you want to keep all the panels straight and all the teeth on the gears. You put your learner behind the wheel of an over-engineered, easy to drive, predictable car like a Volvo. The same goes with a new pet, especially one

Opposite: Fluffy day-old chicks are super-cute but I recommend a few pullets as a better starting point for the first-time chicken-keeper.

such as a chicken that is unlike the cat or dog you might have had previous experience with. For this reason, it is my strong recommendation that you start with point-of-lay pullets of a robust crossbreed or hybrid.

A pullet is a young hen, less than a year old, and 'point-of-lay' means the bird is essentially fully grown but not quite laying yet, or just starting to lay.

These are the best hens to start out with because:

- They are big enough not to be considered a snack by anything with sharp teeth or a sharp beak;
- Being fully fledged, they are much more resistant to temperature fluctuations;
- They tend to be hardy, can be pestered by your kids all day and will still lay an egg the next morning;
- You will get the best laying span out of these hens.

Starting with chicks sounds cool and newly hatched ones are, admittedly, super-cute. On the other hand, they are only cute fluffballs for a couple of weeks before they go into what I call the 'buzzard-stage' of scruffy sprouting feathers.

Also, the mortality rate for chicks is high compared with older birds. So, to be on the safe side, you would normally start out with more chicks than the number of hens you ultimately want. Depending on your luck or skill, you might end up with more or fewer hens at maturity.

My final argument in favour of starting with pullets is that, unless you get professionally sexed chicks from a reputable supplier, you can easily end up raising a flock of cockerels. Cockerels are very tasty, although your kids might not forgive you for serving Harriet-who-turned-out-to-be-Harold for dinner.

At the risk of receiving a nasty letter, I'm also going to recommend you start with a crossbreed layer. Breeders, hear me out! I'm not disparaging purebreeds, just recommending that the first-timer start with a crossbreed gun-layer: a first foray into chickens can sometimes involve a lot of convincing of other family members and plentiful eggs can go a long way towards persuading waverers that the benefits outweigh the misgivings.

43

Opposite: The deep litter floor makes cleaning out this coop easy – simply sweep out and replace. For added insect resistance, throw some lime about before you replace the litter.

Before you start

Mobile coop (chook-tractor)
or fixed coop (aviary)?

Unless you intend to keep your chickens inside your house
(not recommended) the pre-chicken purchase or construction
of a coop is essential. Coops fall into two main categories:
the mobile coop (or chook-tractor) and the fixed coop (aviary).

The mobile coop's main benefit is its mobility, giving the
option of easy relocation. Some even come with wheels. Also, if
you are tentative about your foray into poultry, these are generally
smaller and well suited to starting out with a couple of hens.

Commencing with a fixed coop is a more decisive entry,
but fortune often favours the bold. Setting up a fixed coop will
necessitate consideration of its location and better planning of
how your hens will fit in with the rest of your garden. For more
on the choice of coops go to Chapter 5.

Getting your other pets chicken-ready

The most common misconception I have had to dispel is that if
you have a cat you can't keep chickens – *allow me to retort*! Cats
and chickens are not mutually exclusive (although for every
rule there is an exception: if you own a 20 kg/45 lb Norwegian
Forest cat or an illegally imported leopard, hens might not be for
you). Kidding aside, a fully grown hen has nothing to fear from a
domestic cat and in 15 years of selling chickens I have never heard
of a pet cat killing a hen. (I have had two reports of hens being
killed by feral cats, but they are a different beastie altogether.)

Moggie sitting on the fence, closely observing your flock, tail
swishing with intent, may give you the heebie-jeebies, but will not
impart mortal wounds.

'But my dog is always chasing birds,' I hear you say. There is a
world of difference between a 100 g (3 oz) pigeon and a 2 kg (5 lb)
hen. Also, part of the reason your cat or dog chases these birds is
that they are wild and not part of the household.

All households have their own dynamic, especially where pets
are involved. When you add a new pet you change that dynamic.
A cat knows its place in the family pride – it might think its place
is at the top of the pile, but it knows there is a place for it. Same
goes for dogs: they have a place in the family pack. What you're
aiming for is for the hens to become an accepted part of that pack.

Opposite, top: A silver-
laced Wyandotte and
two Araucanas with their
portable A-frame chook-
tractor home. Below:
Dogs and chickens usually
coexist in harmony.

For this reason, it's best to introduce your new hens to your existing pets. The introduction should be closely observed, especially with dogs, who should be leashed. Existing pets will be naturally curious about the newcomers and we don't want the curiosity to become fatal.

I invented a trick for shaking up the potential predator/prey dynamic at the initial introduction of cats and dogs to new hens. Believe it or not, I got the idea from a hen. When delivering chickens to a customer, I noted the approach of a Doberman (a large and quite intimidating looking dog) from the house. Before I could give the situation a thought, the hen, rather than running away in fright, puffed out all its feathers, stretched its neck up to make it look taller and went for the dog. The fast approach of a were-hen with murderous intent in its beady eyes was too much for the dog, who legged it out of there, double quick. This gave me an idea.

What you do is hold the hen in both hands out in front of you and then run at the dog or cat, squawking loudly like a chicken: *Bekirk! Bekirk! Bekirk!* Chase the dog or cat around the garden. The chicken becomes an unknown. Is it tougher than it looks?

The next step for dogs is to allow them to approach the hen while leashed. Reprimand the dog if it takes an unhealthy interest. Over time, reduce the supervision of the interaction until they can be left alone together. You can tie the dog near where the hens are foraging in the garden or in the hen run; just make sure the leash is short enough that the chickens have dog-free room. The dog might be excitable at first, but the novelty will wear off.

I recommend this as opposed to installing a fence with the dog on one side and the hens on the other. If you do that, you will need to make the fence dog proof, which is a lot more elaborate than a hen-proof fence. The problem with full segregation is that if the hens and dog don't become acclimatised, the dog can see them as prey – then, if they do get together, the dog goes berko and, unless you are quick, the hens can be killed. This can happen if the dog digs its way through the fence, or if the hens flap their way out of the hen run, which occurs occasionally.

Not all breeds of dog are well suited to sharing a garden with chickens. Terrier breeds (fox terriers and Staffies etc) were originally bred for hunting and are harder to keep with chickens because of their natural instincts. That doesn't mean that if you have a terrier you can't keep hens, but it can be a bit harder to make it work.

If it is too much trouble, you can install a fully enclosed hen run, with a roof (either sheeted or wired) and with side fencing dug deep into the ground. It is quite a bit more work, but has the advantage of also being fox proof. This means you can leave the hens on their own for a few days and don't have to be vigilant about the twice-daily routine of locking them in the coop at night and letting them out in the morning.

I've always owned working dogs, which can suffer from boredom. I have no problem with the Border collie herding our hens around the garden. Occasionally, I have observed Freddie-dog take a nip at a hen, especially when I put down the kitchen scraps. However, because the dog interacts with the hens regularly, she doesn't lose control and injure them. Also, if she gets a feather in her mouth, she has a devil of a time spitting it out!

The mention of kitchen scraps brings me to another point of consideration for dogs and hens. It is probably best to restrict the dog's access to hens' kitchen scraps as no dog I know has ever shown self-control where food is concerned. Unless you want your pooch to become a hairy balloon, keep the hound away until the hens have got through the best bits.

47

The final word on dogs is that in 15 years of selling chickens, and over 4000 chicken coops sold, I have only had six coops returned because people couldn't make chickens work with their dog. Three families had terriers and one had a little yappity thing that wouldn't stop barking. So, although I can't tell you the proportion of households with dogs that would be incompatible with chickens, it's obviously a very small one.

For cats, let the cat see you holding the hens and interacting with them. This helps the cat understand that the hens are welcome. Whether his or her highness cares is another matter.

Chickens and rabbits get along fine and I assume the same goes for gerbils and guinea pigs. However, because the hens need outside ranging during the day, you need to leave the coop door open (unless your coop has an enclosed hen run). In my experience, most people keep their rabbits in their hutch all day with the door closed, but if you have a rabbit-enclosed garden and allow your rabbit to free range, they can share a coop/hutch as well as the garden.

Getting your garden chicken-ready

As a general rule, hens will peck at and eat anything soft-leafed and palatable. The bad news is that puts your vegetable garden and herb pots in the firing line. The good news is that many weeds are soft-leafed and at risk of consumption, too. In fact, I have had customers buy chickens for the specific purpose of controlling weeds in an unloved part of their garden.

How do you stop hens from eating your herbs and vegies? You can't. The only solution is to prevent the hens having access to them by fencing either the vegies or the hens.

'The lawn! The lawn! Will the hens destroy my lawn?' This is a classic length-of-string type of question and I get asked it all the time. The number of turf-proud home-owners and the level of their passion never ceases to amaze me.

It also often seems to be a deal breaker in intra-marital poultry-keeping negotiations, so I will be quite expansive on the subject. The simplest solution is to install a hen run (have I mentioned that before?). If a hen run is not an option then bringing hens into your household doesn't necessarily condemn your mown paradise into a re-creation of Death Valley.

Hens tend to peck the new growth of grass, which I assume is tastier and juicier, and that means they can be used as feathered mowers. Just don't expect them to keep your lawn looking like a lawn bowls green: the trimming is rather more mottled and haphazard than that.

You also need to ensure that once the new growth is adequately pecked you get them off the lawn, or they will start in at the roots and runners and damage the grass.

If the hens are going to have permanent access to the lawn, there are a few parameters you can consider in predicting whether the lawn will survive. How much sunlight does the lawn get? How big is the lawn? Is there any other soft-leafed fodder accessible to the hens? How many hens do you ultimately want to keep?

Too many hens on a small, shaded lawn and they will denude it completely. Feathered Desert Storm – mission accomplished! A few hens on a large, sunny lawn and you will hardly notice their presence. You see why it is a length-of-string question? Also of consideration is the seasonally varying growth rate of grass: a lawn that has shown no impact of poultry pruning in spring and summer might need to be protected from pecking in winter.

Opposite: With plenty of sun and a well-established lawn, the impact of chickens on your grass should be negligible.

Before you start

Chickens like to take dirt baths. Flapping dust and sand through their feathers removes dead skin cells and whatever other gunk resides in there (I don't like to think too much about that). So, hens will make a small depression somewhere within their range to take a wash in. This can be in the lawn, but is usually somewhere more weather protected and drier, under the canopy of a tree or shrub. The good news is that they rarely establish more than one dirt bath unless you have lots of hens, so you don't need to fear a lunar-esque lawnscape.

As for the rest of the garden, there are two main things to consider: root disturbance and mulch relocation. Hens will scratch the soil quite vigorously, looking for ground-dwelling insects and seeds, anything edible really. Hens are omnivorous, so anything that moves (and much that doesn't) is on the menu.

Hens don't dig holes like dogs or rabbits and are definitely not adept at tunnelling with a spoon as portrayed in *Chicken Run*. Their scratching, however, can be detrimental to shrubs and trees, especially if these are shallow rooted and/or sensitive to root disturbance. Once again, if you have a few hens in a big lush garden with plenty of space and forage, this won't be a problem. But with more concentrated numbers or space, scratching can become an issue, especially in a hen run.

Fortuitously, the solution is simple. Just arrange some bricks or stones in a circle around the base of the plant at a distance where root disturbance could cause harm. I've also used that green plastic mesh with 5 cm (2 inch) squares that you can get from the hardware store. It is easy to cut, is cheap, lasts forever and looks less industrial than wire mesh. Just cut a square the size of the area you need to protect, from the centre of one side cut halfway in, and then in the middle cut out enough squares for the trunk of the tree or shrub.

The other annoyance in relation to hens and gardens is the kicking around of mulch. I recall one garden where there was no mulch left on the garden beds: it was all spread over the lawn avenues in between. The owners were raking it back daily and it was driving them nuts. The solution was, you guessed it, a hen run. Mulch hides all sorts of yumminess for hens and they can be determined when investigating it. Making a lip around the edge of mulched beds and planting understory shrubbery will reduce the kicking of mulch onto paths.

Opposite, clockwise from top left: Wrapping vegies will protect both roots and foliage; Hens can till a fair amount of soil looking for food; At the end of the season, chickens can be put to work to clean weeds and seeds from the vegie patch.

And, in case you were about to ask, I'm not a subscriber to the school of thought that says that bantams are better for your garden than full-size hens. Bantams are just a diminutive version of a bird with the same breed characteristics. Being smaller doesn't necessarily mean that they cause less damage to the garden. (The same goes for my youngest boy, funnily enough.) In my experience there is much more variation between breeds and individuals and any generalisation about bantams doesn't hold true.

Truth to tell, there are some breeds that have a reputation for being gentler to gardens: Silkies, for example. However, I have the sneaking suspicion that the breeds that are reputed to be gentler have an accompanying side effect of being rubbish layers. This is because laying hens have a huge metabolic commitment to egg production and your gun layer is constantly hungry, incessantly scratching to find the next morsel.

Also, I'm not a huge fan of bantam eggs: too small. Someone of my robust physique can't derive sufficient nourishment from a teensy bantam egg. I'll leave those to the skinny hipsters.

So, to wrap up this section: chickens and gardens *are* compatible, the impacts can be managed, and more hens and/or less space will require more effort.

Rules and regulations

Even back when I started Rentachook there wasn't a single local authority I ever came across that prohibited the keeping of a small well-managed urban flock of hens. These days, local councils are much more likely to encourage than discourage the keeping of hens as they've been recognised as a sustainable pet.

That being said, there are regulations to comply with and your local authority is the place to enquire about specifics. In general, the requirements are similar in all urban jurisdictions. As you move out of the suburbs into the rural fringe, even these restrictions tend to fall away and the focus changes to regulating farming practices for poultry farms.

Regulations for urban hen-keeping focus on keeping the hens in such a way that they don't cause a nuisance to your neighbours. As a general rule, regulators (council health officers etc) will only get involved if they receive a complaint and, even

Opposite: The deep litter mulch in the hen run gives the chickens plenty of entertainment looking for food, and is easy to manage because they bury their own poo.

then, will only take an interest if your hens are too numerous or poorly managed. One way to head off a visit from the fun-police is to have a chat about your idea of keeping hens with the neighbours beforehand. Let them know that if they have any concerns, they should let you know first so you can sort it out. **There are some regulatory requirements that seem to be replicated by most councils, including:**

- **Roosters are prohibited**
  Because they crow and it is loud. Understandably, as we now all live far too close to one another (and have little tolerance for loud crowing, especially in the early morning). Not having roosters does not affect the hens' laying: it just means the eggs are infertile and cannot turn into chicks. I have recently come across a new product called a 'cock collar' (*now, I know what you're thinking, but please stop*). The collar is placed around the neck of the cockerel or rooster and it stops it releasing the inhaled air all in one go. This means that the rooster still crows but at much lower volume. I have not had the chance to personally verify its efficacy, but the principle is sound and it might mean that cockerels will be able to grace the city or suburban garden again in future.

- **There is no limit on chicken numbers (in most council areas) but they must be kept in a healthy environment**
  This just means you have to look after them and keep them clean – read on. However, those areas that do have limits set it at 10 or 20 or, very occasionally, as few as five, meaning that you can still keep a good-sized flock even in the most restrictive of areas. Keeping them clean also entails preventing or managing vermin, but there is a whole chapter on that later, so you'll be fine.

- **They must be kept in a suitable coop that is of adequate size**
  The birds must be housed in a suitable coop (the term 'suitable' is very loose, meaning not too small for the number of birds, and able to be kept reasonably clean). It also means that it is permitted for them to free range in your yard as long as they are prevented from escape in some way. This usually means having an enclosed yard and clipping their wings if the birds are flighty or prone to escaping.

55

Opposite: Steve, a proud Araucana rooster. Steve is prohibited in surburbia because of the noise he makes.

Before you start

When I say an enclosed yard, really the only thing that matters is that you get along well enough with your neighbours that they won't go ballistic if a hen should escape into their garden.

 The coop must be a minimum of 'a certain distance' from any dwelling

I have been told the distance from a dwelling rules relate to the old practice of people building a fixed chicken coop along their fence line in the backyard. You can imagine that if such a coop had lots of birds, was not maintained and was located right next to the neighbour's kitchen window, there could be a hullabaloo (hence the distance rule). Where the distance rule comes in to play, there can be some debate as to whether this also applies to the hen-keeper's own dwelling. Not being a lawyer, I cannot give advice on such an interpretation. However, it seems unlikely that you would be seeking your regulator to enforce a requirement that applied to your own pets, no?

So, in summary, keeping hens, even in a small garden, isn't going to draw the ire of the constabulary, so long as you manage them properly.

Getting ready for the (minimal) daily workload of chicken care

I cannot state often enough that if you are spending the dark side of 10 minutes a day in poultry management tasks, you are probably overdoing it.

Morning: Amble down to the coop and say 'hello ladies'. Let them out. Check they have plenty of water and the feeder is topped up. Depending on what type of feeder/waterer you use, there may be no daily work here. This is also the best time to give the hens kitchen scraps as they will have the whole day to pick over them. Collect the eggs!

Evening: Some time between dinner and bedtime, mosey on down and shut the chickens in.

It is my strong preference and recommendation that hens are out of the coop, either free ranging or in the hen run, all day. It is the best for them and reduces the workload (the more they are confined, the greater frequency of coop cleaning required).

Opposite: A cubby house converted into a chicken coop. Take care with older cubby houses to ensure they aren't constructed from CCA treated timber – it isn't advisable for this to come into contact with food such as eggs. Overleaf: Two magnificent roosters – an Australorp, on the left, and a Rhode Island Red.

Periodic cleaning: Another length-of-string question I get asked is 'how often do I need to clean out the coop?' The answer to all such questions is, of course, 'that depends'. When asked to elaborate, I'm usually much more helpful and reply 'when it gets manky'.

There are simply too many variables to answer such a question with a frequency-of-cleaning reply. This seems to be the response that routine-orientated people (quite unlike me) seem to be seeking.

The more hens you have and the smaller the area they are kept in, the more often you have to clean it. This is also dependent on weather, as accumulated chicken poo is much more likely to create a miasma if it is hot and wet rather than cold and dry.

Experience with your own flock and coop will tell you the point at which, if left, it will start to smell. Also, there is a tipping point beyond which the task of cleaning gets harder than it should be because you have left it too long.

So, if pressed, I'd say, with seasonal variations, that if you are spending much more than an hour a month you are working too hard and need to take a look at your set-up. More frequent cleanings will also entail less time and effort, so, if you were to spend 15 minutes every couple of weeks say, your overall cleaning time would, in all likelihood, be less than the hour a month. Please jump to Chapter 5, the Housing section, for more on coops and their management.

And when you go away on holidays?

The millstone of pet-owners… So, you can never leave town again? Well, not quite, with chickens.

If you are going for a short trip, say a long weekend, you can keep them locked in the coop. Now, before you remind me of my oft-repeated mantra that hens should not be kept cooped up, there is a good reason for them to stay in the coop if there is no-one around to lock them up at night.

In the coop they are safe from predators and, while they won't love you for it, in the long run it's preferable to be confined than eaten. I've seen hens rattle their beaks along the wire in the same way a prisoner might run a metal cup across the bars – but don't let them play the guilt card.

Before departing for your weekend there are some precautions to take. Leave the hens plenty of feed but, more importantly, ensure that there are at least two separate sources of water. If the hens knock over a feeder they will eat the feed off the floor, no problem. But if they spill the water, it will drain away and, in hot weather, 24 hours without water will kill a hen.

Speaking of hot weather, also make sure that the coop is placed in a shaded location or otherwise sheltered from the midday and afternoon sun. Don't forget that hens are wearing their own personal feather duvets.

If your coop has a floor covered in shavings or a dirt-floored enclosed hen run, you might also consider spreading and hiding some whole seeds to stave off boredom (the hens', not your own). If you've got a shavings floor it would be nice if you replaced the shavings before doing this.

If your coop doesn't allow you to strew seeds and hide them, you can put the seeds in a thin sock or stocking and hang it up for the hens to peck at. Hanging fruit or a cob of corn in the coop can also mitigate the baleful look your hens will give you for leaving them locked up.

The gold standard in vacation care is to have someone let the hens out in the morning, check their feed and water, and lock them up each night. I find that bribery is effective in eliciting the support of friends or neighbours: keep an eye on my hens and any eggs you find are yours. A cunning plan is to soften up your neighbours with a few irregularly supplied six packs of eggs beforehand. Home-grown eggs punch well above their weight in the gift stakes. Once you have friends and neighbours well and truly hooked on proper eggs, it is not much of an extension to request they come and get the eggs themselves each day for a short period (and look in on the hens). The absolute minimum care your hens will need is someone to check they have sufficient food and water every three days.

If you're going away for a few weeks and don't have someone to let them out and lock them up each day, a lesser option can be to let them out one day in three. This way they get some outside time but it's not as demanding on your helper.

(In the early days of Rentachook one of my customers went away for five weeks to the UK. I suggested a neighbour to chook-sit and all went well. Too well, in fact, because when the owners

got back and resumed care of their hens – and consumption of the eggs – the helpers were bereft. Solution: buy another coop. A few months later both families had coinciding holidays and engaged another neighbour from across the street. Same again: upon return, another coop sold. I was starting to rub my hands together: three coops in the one street! Soon everyone would have one – I had visions of private jets and luxury chook-mobiles. Alas, it was not to be.)

An alternative, if you have a portable coop, is to transport the chickens, coop and all, to your helper's place for the duration of your trip. There are also services where people will come to your house to water your plants, walk your dog etc, and many of these will include simple chicken management. Don't forget that you will have to pay them for two visits on any day the hens are let out, so they can be locked up as well.

One final option is to add your hens to the flock of a friend. There will be the inevitable rivalry as the hens establish a new pecking order that incorporates the newcomers. However, they will all be adult and you are adding multiple birds, so the likelihood of problems is minimal.

63

Opposite: Sugar cane mulch and pea straw are good for the floor of a fixed coop if you can keep it dry.

Before you start

4

CHOOSING AND COLLECTING YOUR HENS

How many chickens to start with?

Good question that one. I retort with: 'How many eggs would you eat in a week?' This is going to go in circles if you reply 'lots'. Think about this: a healthy young crossbreed hen will lay an egg a day if fed correctly. That means a dozen eggs a week, give or take, from the minimum flock size of two hens.

If you consistently eat more than a dozen eggs a week then I would say start with three hens and, if your mob monsters eggs voraciously, then perhaps even four. Remember, if you don't have eggs for breakfast tomorrow, your lovely ladies will still present you with your golden presents and this can add up quite quickly. Never fear, you will have no trouble finding eager recipients for your excess.

(While I think of it, I keep hearing that people have read on the Internet that you shouldn't have three hens, or odd numbers of hens. What bunkum!)

So, why wouldn't you just start with lots of hens and make lots of friends with all the eggs you'll have? Well, for two reasons.

Firstly, keeping hens is a new venture and the more hens you start with, the more likely you are to face challenges that make

Opposite: A flock of Araucanas. Note the woody mulch – the woodier the better, because it lasts longer.

your chicken-keeping less wondrous than it might have been. Best to start out slow and grow into your new pastime.

Secondly, if you fill your coop with young hens at the get-go, all your hens will be the same age and there won't be room for any more. 'Who cares?' you say. 'You will,' I reply, because your hens will all age together. Which means that at the beginning you will have so many eggs you won't know what to do with them all: a boom situation. Then, in time, you will end up with a whole flock of hens who are ageing past their laying prime: a bust situation.

This happened to me a few years ago when I got into the bad habit of accepting older hens from customers who no longer wanted them. (*Big softie, I was… not any more, I'm as tough as cold butter now.*) I had a flock of a dozen or so but was only getting one or two eggs a day, as the average age was about seven years.

For this reason I recommend you keep what I call a 'rolling flock'. A rolling flock is when you even out the boom and bust by starting with just enough hens to meet your egg needs and then add new hens when the quantity of eggs drops as the hens age.

Say you have a coop that is capable of comfortably holding six hens, but your family loves their eggs and a dozen a week is never quite enough: start with three hens. Then when, in a few years, your hens are no longer laying daily and the number of eggs drops, add two more hens to your flock. Plenty of eggs again, amply making up for the pending happy retirement of your original hens.

By the time you have to do this a second time round, it is likely that one or more of your original flock will have fallen off the perch and your coop will happily accommodate the new additions. In this way you keep a rolling flock of between three and five hens indefinitely and are never short of eggs. Clever, eh?

Buying healthy hens

I've bought my fair share of dodgy hens over the years – and I'm not talking about back-alley deals with trench-coat-wearing ne'er-do-wells. There can be a bit more than you imagine to buying healthy hens.

Hens can get parasites and diseases that spread quickly through large flocks held in close confinement, such as are housed at large breeders or commercial rearing farms. Conditions such as chronic respiratory disease are also notoriously hard to eradicate and involve emptying sheds and cages of birds and thoroughly disinfecting. This means a lot of work to do it properly but, if you don't, the next batch of hens can catch it. Everyone who sells hens for any length of time gets these problems.

Accordingly, for the unscrupulous, foisting an unwell hen on the unsuspecting can make good financial sense as it gets rid of a problem and completes a transaction. In my business I always have an ongoing relationship with my customers so, if I were to sell hens with problems, I would end up having to nurse the customers through the issues. Easier for me to keep the hens and solve the issues myself.

So, the first recommendation I give you for buying good hens is to buy them from someone who is likely to want repeat business from you. For this reason, I would avoid buying hens from rural stock auctions until you are more experienced.

I pay a lot of attention to the feel of the place I buy hens from. Have your eyes open to the environment and the signs of care, or lack thereof, of the hens. Don't go forgetting that if it is a commercial operation, the hens will be treated as a commodity. There is nothing wrong with this, but an evident lack of care for their wellbeing is indicative of an increased likelihood of their having health problems.

Look at the chickens themselves. Do they look bright-eyed and bushy tailed? By which I mean do they look sharp, engaged, lively? Or do they look dopey and sluggish? Two other things I like to check when assessing a new hen supplier: are there any respiratory ailments or mites/lice?

Respiratory ailments are relatively easy to spot as the hens will display symptoms similar to those of humans with a cold: snotty noses (note the nostrils of a hen are the two holes in the top beak) and wheezy or rattly breath. They can sneeze, too. One or two sneezes can indicate nothing at all, but if the whole flock is sneezing...

Mites and lice show up as little black or white dots in the feathers and, if you suspect you are seeing them, ask the seller about it. When they are really bad your arms will feel itchy after handling the chickens. See Chapter 9 for more on mites and other ailments and how to treat them.

As a general rule, I won't buy hens from a supplier for at

69

least a year after being supplied with hens with problems. The problems listed above can all be managed, but they are a pain. As a retail customer you are best looking elsewhere for your hens. Let your wallet do the talking.

How old should my chickens be?

The best hen for a first-timer is a robust crossbreed layer at point of lay (16 to 20 weeks old). As your experience increases, you can then take a punt on some of the gorgeous purebreds around. (*There's nothing like a fancy chicken breed to bring a smile to the face or start a conversation.*)

To make a good guess at the age of the bird, there are a few telltale signs. The comb is a good starting point. The comb is the bit on top that you mimic when you put your hand on your head, fingers upward and spread, when you are imitating a chicken. (*Come on, don't deny it, we've all pretended to be a chicken at least once.*)

The comb is a good indicator of maturity, growing and filling out as the hen matures toward laying. (With every rule there is an exception and there are some breeds with what is called a 'rose comb', where this isn't the case.)

If the comb is small and the hen looks slighter than you would expect for the breed or variety, then it is likely that the hen is young, perhaps 12 or 14 weeks. There is an incentive for poultry-sellers to suggest their young hens are older than they are – they make more from selling you hens at a younger age because they don't have to feed them for as long.

If these are your first hens, this is no big deal really; you will just be waiting longer for your first eggs. However, if you're adding younger birds to an existing flock, they are likely to get bullied and you might need to keep them separate until they are a bit older and wiser.

On the other side of the coin, you don't want to be buying hens that are old, especially if you have paid for and are expecting young hens. Commercial egg farms turn over their flocks at 12 to 18 months old to maintain the highest possible productivity. For the unwary, you could end up buying ex-commercial layers sold as 'point of lay'. The downside of this being, obviously, that you miss out on the birds' period of best laying.

Opposite: An immature Leghorn pullet. Note her small sawtooth comb, compared to the mature Leghorn on page 11.

This brings me to the 'saving' of ex-battery hens and giving them another life in retirement in your backyard. You'd expect me to be bang up for that, right? Nope. At the risk of attracting online vitriol, I just can't see the point.

Really, what difference is the saving of five ex-cage hens out of a shed of 100,000? Also, because the hens live in these abominable conditions, they often have permanent mobility problems. Another thought-bomb is that the ex-cage hens are essentially a waste product of factory farming. Is the purchase of such hens not counterproductive to the intended outcome of stopping such farming techniques? Over to you.

As hens age, their feathers become more ragged. If you have ever seen an emu feather up close, it's a bit like that but less pronounced. So, if the ISA Brown or other rust-coloured crossbreed hens you are looking to purchase have ragged feathers, check for young-looking hens in the flock, or for batches of hens of different ages at the farm. If you see neither, these could be ex-commercial hens which are cheaper to buy than young growing hens.

Transporting chickens

I've often been asked on the phone: 'How do I transport the chook home? What do I bring?'

Dave: 'Bring duct-tape.'

Customer: 'Duct-tape?'

Dave: 'Yep.'

Customer: 'Why?'

Dave: 'Then you can tape the chook to the bonnet of your car, as a hood ornament. Kind of like a Rolls Royce, but with less of an angel and more... well... poultry.'

Customer: 'Really?'

Dave: 'Nope.'

Sometimes I can't help myself.

Chickens are best transported in a plastic cat carrier if you have one. These are great because they are easy to hose chicken poo out of if you forget to put in straw or newspaper. They also have plenty of ventilation to keep the hens cool.

Keeping them cool is very important when transporting chickens, especially in summer. So, don't make a side trip to the

Opposite: A mature ISA Brown with large comb. Her slightly ragged feathers give away that this is an older bird and not point of lay.

pub on your way home, leaving them in the car. If a pub stop is mandatory, take them in with you – it's a good way to start a conversation. (Embarrassingly, I have to confess that I have actually done this.)

You might even find an egg in the carrier after the trip home if the hens you've bought are already laying. This has happened for me many times and I have even been known to surreptitiously sneak an egg into the transport box when children have come along for the pick up of hens (*shhh... our secret*).

If you don't have a cat carrier, a cardboard box is fine. A 12-bottle wine carton works well for two hens; they travel well together. Air holes won't suffice in the carton, you need some ventilation – remember the overheating danger. Cut a 2–3 cm (1 inch) wide strip, around 20 cm (8 inches) long, out of two opposite sides of the carton – the box should still be rigid enough, but will get some cross ventilation. The hens can also poke their heads out, which kids love when they ride shotgun or on their laps.

The old-school trick of tying their legs together and laying them in the back of the car, or transporting them in a hessian sack, is *soooo* last century (and really not very nice).

Minimum equipment requirements vs gearfreak

With any hobby or pastime there will come a plethora of unnecessary accoutrements available for purchase. If you feel the need to reward yourself with some retail-therapy, by all means, go right ahead. The Internet is at your fingertips: go to town!

At one end of the spectrum, it is possible to buy nappies for hens so that they can have full run of the house without leaving a trail of excreta. At the other end, feeders made from recycled plastic milk bottles last pretty much forever and can be made so that no feed is spilt or wasted.

Predictably, I sit at the milk bottle end of the spectrum and will use this section to outline the minimum you need. Where you go from there is up to you. Don't get me wrong: buying groovy stuff is fun and I have a great penchant for elegantly designed equipment that is supremely fit-for-purpose. I also encourage you to design and make your own equipment for your hens. There is a long tradition of home-made coops and ingenious equipment for chickens; it is also a great way to engage children.

Opposite: Two hens will travel happily in a cardboard box. Two long cut-outs, on opposite sides of the box, are better than airholes for preventing overheating.

As a minimum, you will need:

- ⵏⵏ **Coop**
  It will need to be fox proof. Even if you are in an area that is fox free, New Zealand for example, there are other predators that take their place. Hens need a predator-safe place to roost at night. See Chapter 5 for full information on coops.

- ⵏⵏ **Feeder and waterer**
  There are many versions of feeders, ranging from hen-operated, self-closing jobs to a dish on the ground. The main priority for a feeder in my opinion is that it allows the hens to eat without letting them spill feed. Waterers can range from continuously replenishing devices filled from the tap or a tank, to the above-mentioned milk bottle with a hole cut in the side (see page 111).

- ⵏⵏ **Nest or nesting area**
  The coop needs to incorporate some suitable nesting material and the area be preferably dim or dark. Most coops made for hens will already feature a nest, so all you need add is the nesting material.

- ⵏⵏ **Dry storage for nesting material**
  This may seem obvious but I am trying to include the absolute minimum and stored nesting material will rot if it gets wet. Also, don't let stored nesting material become a favoured home location for rodentia.

- ⵏⵏ **Rodent-proof storage for feed**
  A rubbish bin with a lid will do. I have 80-litre (20-gallon) open-top plastic drums that were originally used to import sardines, but anything that keeps Jerry out will do, especially if you don't have a Tom.

The first few days–weeks–months

You pull up in the driveway with a box of hens. Kids, if you have them, going nuts and jumping out of their skin with excitement; your partner, or yourself, doing the same. I have seen people from the ages of two to 80 hop up and down on the spot, clapping their hands with glee at the arrival of their first hens. Cherish the moment; it is a great memory. Take some time to enjoy them before their novelty wears off. They will always be fun to watch,

Opposite: A day-old chick. This will be a cute fluffball for about a week before it begins to grow feathers at the wings and tail.

but many customers have described that first afternoon, kids enraptured, glass of wine in hand, as being special. To quote Ferris Bueller: 'Life moves pretty fast; if you don't look around once in a while, you could miss it.'

The easiest way to get the hens used to their coop and new environs and to get them to return to the coop at night is to put them in the coop straightaway and leave them there for the next couple of days. This, however, doesn't take into account the insane urge to play with your new, feathered toys.

This urge need not be repressed, especially if kids are involved. If this is the case, release them into the coop but leave the door open. There is no harm done if you feel the need to hold your hens before releasing them, as well.

Start by putting them in the coop and make sure they have water and food and know where to find it (see page 108). If the hens are to be released for play on the first day, give them a little while to gather their thought (I don't believe hens are capable of more than one thought). Not being a hen, I have no idea how stressful a car trip and new environs are for them. Food, water and a short respite are wonderfully restorative either way.

I encourage you to clip the hens' wings before they are let out. A happy day could turn into a sad one if your new pets fly over the fence and are lost. There is more information on clipping wings on page 139.

If you are going to let the hens out on their first day and you have other pets, it is probably easiest to leave those pets in the house for now.

At dusk, if the chickens are still out, you should herd them into the coop. You don't necessarily need to catch them; shooing them in is fine. I find it is easiest with a broomstick, tomato stake or similar in each hand so you have multi-directional shooing control. If you have left it too late and they find their own roost for the night you will have to find and extract them or this may lead to roosting outside the coop in future.

Once they have got used to the coop as their night-time roosting locale, they will put themselves to bed each evening. This can take a few days. If you find them perching on the ground or, worse still, in the nest, wait until it is darker and they are dopey and pick them up and put them on the perch. No harm will come to a hen that perches on the ground in a properly predator-proof coop and some hens never work out the perching thing, but it is worth giving it a go.

You will most likely have to wait until the hens settle and are fully mature before you get your first eggs. Don't freak out if you get a few eggs in the first few days and then the hens stop laying. I have had to use my special calm voice on a number of occasions where panicking customers have thought the laying has stopped due to some omission of theirs.

What has happened is that the hen had a number of eggs in the pipeline and nothing was going to stop them coming out. However, the change in environs, change in diet and/or the stress of relocation has caused her to put her laying on hold. When equilibrium returns, so will the eggs.

For those who bought younger or not fully mature hens, there will be a wait. There are two ways to approach the recurring disappointment of going to the coop in the morning and returning empty handed. One approach is to curse the hens for their failings and begin to doubt the wisdom of your purchase. This is much more likely if nagging children add their special layer to the experience. Alternatively, buy a cheap calendar, if you don't already have one, and everyone in the household gets to nominate dates for the arrival of the first egg. The winner of the egg-lottery gets the egg for breakfast, of course.

Eeurgh! The first egg is all misshapen! Or, the first eggs are tiny, elongated, have crusty bits on the shell or no shell at all. Your partner says: 'I told you not to get the hen with the beady eyes and evil look; she's turned out to be a Frankenhen. Quick, grab the egg and destroy it before nightfall!'

Remain calm. Imagine that inside a hen is a little production line with many stages that need to be successfully completed for you to get an egg that you would recognise. Unsurprisingly, it is not uncommon for the first few cars coming off the production line to have a wheel missing. Don't worry about it, she'll sort herself out.

Over the next few months you and the hens will get to know each other and fall into a routine. These are the halcyon days of hen keeping: novelty, plenty of eggs and generally few, if any, issues to resolve. You are a happy hen owner now.

Smile, people: it helps dry out your teeth.

5

THE CHICKEN HOUSE

The minimum requirements

It is possible to keep chickens without any form of housing, and in Southeast Asia there are still wild birds that closely resemble the ancestors of your modern hen. (*Nimble little things they are too: they can fly quite well and are quick on the legs. Goodness only knows how we humans managed to catch enough of them to domesticate chickens in the first place.*)

The point of this waffle is to explain that your modern hen is much bigger, slower and more productive than its wild forebears. Consequently, they need protection from predators: they are just too easy to catch. And, while you're building something to protect them from being eaten, you might as well provide protection from adverse weather, give them somewhere to sleep and make the eggs easy to find to boot. So, that's the minimum requirements for a coop right there.

Fox proof

Opposite: An elegantly designed and beautifully made coop is a great addition to any garden.

Backyard Chickens

The most common predator to hens worldwide is Mr Fox. (I am sure Mrs Fox must exist, but I always see my arch nemesis as male.) If you can protect hens from his wily, vulpine attentions, you can protect them from most other predators. Although I have

a vexed relationship with foxes, I find them beautiful, intriguing adversaries, yet treacherous. To keep Mr Fox out, you need to completely enclose the hens. They need walls, roof and floor – kind of like a panic room. With the door closed, our VIPs (Very Important Poultry) can await the coming of the cavalry in safety.

Don't underestimate Mr Fox. Chicken wire will not be sufficient: a fox can chew through it. You need welded aviary mesh. Whether your coop is mobile or fixed, you will need a way to stop him digging his way under the walls. For a mobile coop that means a mesh floor. For a fixed coop, that is easy: have a hard floor. Concrete, bricks or pavers will do. Failing that (especially if you have loose or sandy soil), dig a trench around the outside of the coop and bury mesh to a spade's depth. Alternatively, lay mesh on the ground from the base of the walls. The mesh should extend 40 cm (15 inches) or so from the base of the wall; the distance is not critical but foxes will try to dig under at the base of the wall and the mesh is to confound them.

In 15 years or so of manufacturing portable chicken coops and selling them all over the place, I have only once had a hen killed by a fox when it was inside the coop with the door securely fastened. This was in sandy soil on the coast and the fox dug a trench right under the coop and bit the legs off a hen that was roosting on the ground. You have been duly warned.

Weather protection

With the exception of Silkies, chickens are reasonably waterproof. Not like ducks, mind you, but to a moderate extent rain will slide off them. Hens need to be dry and undercover overnight and have access to shelter during the day. I find some hens will scuttle back to the coop at the hint of a sprinkle, while others will hang out in a downpour with not a care in the world. If your hen run has shrubbery or overhanging trees, most hens will congregate under these when it is raining or stinking hot.

Speaking of stinking hot, heat really knocks the chickens about much more than the cold. If you have options for siting the coop and hen run, I strongly recommend that you ensure they have shade in the afternoon. Trees and other planting are best for this, but you can also use the shaded side of a building or fence and as a last resort, shade cloth. (I go into heat management of birds in more detail in Chapter 9.)

It is the overnight heat management of the coop that is most important. Chickens are little heat beads with a normal body temperature around 41°C (106°F), to as high as 45°C (113°F). Now, imagine you're wearing a thick feather duvet, it's 40°C (104°F) during the day and still 25°C (77°F) at midnight and you're trying to get to sleep. During extended heatwaves I have had customers buying replacement hens for ones that have been found dead in the morning from heat exhaustion – just 'fell off the perch'.

This is a regular problem with a certain type of coop that has poor ventilation in the roosting area. The heat the birds generate just can't dissipate and they bake themselves. If you live in a climate with hot nights, you want plenty of airspace above the perches and at least one side of the coop open with mesh to allow airflow. Open and airy, that's the trick.

Of course, this is the antithesis of what you need in extreme cold. Draught management is your focus then. (I'm talking snowy winters here, folks, not the sort of namby-pamby cold that some of my fellow Aussies complain about, when it's cool enough to put on a whole second layer over your t-shirt.)

Chickens are remarkably well suited to the cold. When they perch, they cover their legs with their feathers, insulating them. In fact, in still-air conditions it needs to get well below zero to bother a chicken; cold enough to freeze the comb on top of their head and give them frostbite. In these conditions, perches closer to the roof allow the radiated heat from the hens to be contained around them. You can also fashion a hood over the perches to trap air around the hens, but this would only be warranted in really properly cold, above-the-snowline areas.

If your hen does get a frozen comb, defrost it slowly by pouring cool and then warm (never hot) water over the comb. Once the comb is thawed, apply petroleum jelly (Vaseline).

Cold, moving air is bad for chickens though, so you need to prevent draughts in the roosting area. Don't go too far and make the coop airtight, or moisture will be trapped, condense, promote mould growth and make the hens sick. Also, frostbite occurs when you get cold and moisture – the intent of the ventilation is to remove the moist air, even if it means some loss of heat in the coop overall.

For mobile coops, enclosed hen runs or coops with mesh walls, draughts can easily be managed using a tarpaulin or heavy-duty builder's plastic. Peg it down or pin it onto the coop so that winds don't lift it away. Timber cover strips screwed to the coop through the plastic work well.

Moisture can also be managed within the coop by regular removal of poo from the floor and under the perches and, in a fixed coop with a hard floor, by making sure the floorcovering material is not damp. In these conditions, floorcovering such as wood shavings can be swapped for sand, which evaporates moisture more readily.

Provides somewhere to sleep

Roost or perch: whichever you prefer. Just a quick reminder here: 'roosting' is sleeping, most commonly done on a perch; 'nesting' is laying eggs in a place. However, some hens are too dopey to understand there's a difference and they go into the nest at night to roost. Meaning that they poo in the nest overnight, which is very annoying and gets poo on the eggs. For more on this go to *Roosting in the Nest* on page 142.

Think of chickens as slow, fat descendants of tree-dwelling wild birds. Yep, they like to sleep in trees (or on branches to be more accurate). A branch will do fine but construction timber is easier. For my perches I use 90 x 35 mm framing pine (nominally 'two by four'), split down the middle. This leaves a 35 x 42 mm rectangular section. Not that the actual width (35 mm or 1–2 inches) is critical, but it just can't be too narrow or the hens wrap their feet right around it and get foot cramps. As a rough guide, the perch definitely needs to be wider than a broom handle.

The next element to consider with perches is to stagger them like the rungs of a ladder leaning on a wall. Hens poo throughout the night, and if you have perches vertically above one another the Top Hens will poo on the Nerd Birds. It doesn't matter if the perches aren't completely level either.

Some hens don't naturally get the hang of perching and they roost on the floor. Don't fret too much over this. If you see it happening, you can pick them up and put them on the perch, but if that doesn't work just let it go.

Opposite: An excellent perch – broad enough that the hen isn't wrapping her feet around it and getting cramp.

Provides a nest for eggs

It is possible to get too excited about the requirements for a nest and fuss over detail. There are some firmly held beliefs out there about the number of nests needed, the size of each nest, the elevation, whatever. Don't overthink it; the stress will send you to an early grave. If you don't provide a nest at all, the hens will just make their own – the problem being that its location will be convenient for their needs, rather than yours.

I have seen excellent nests made from 20-litre (5-gallon) plastic drums, the catchers from old lawnmowers, wooden wine crates, all sorts of things. Hens will share a nest: you don't need to provide one for each hen. I also find that, even if there are many nests available, most of the hens will use just the one. They growl at each other if two want to use it at the same time, although some will lay together at the same time.

Hens like to hop up into a nest that is a foot or two (30–60 cm) from the ground. The height of the nest box in my aviary-style coop at home is 40 cm (15 inches), but it is not critical. Nests on the ground are fine, too, and the coop I designed for Rentachook has the nest on the ground floor to simplify the design and save costs. It is elevated slightly to keep it off the ground and dry. See page 107 for more on nests.

Types of coop

Now you have an idea of the minimum requirements, anything that meets those requirements is a coop. I definitely don't want to stifle your creative spirit, and typing the words 'chicken coop pictures' into your favourite search engine will show you what I mean. I am particularly fond of converted cars as coops (just add the word 'car' to your search).

For convenience, I have decided to divide coops into just two types: the fixed coop and the mobile coop or chicken–tractor. (*Oh, and before we go any further, the 'car chicken coop' is in the 'fixed coop' category – my book, my rules.*)

Fixed coops

Pretty much everyone I know who keeps chickens long-term eventually gravitates towards the fixed coop. Having said this, I'm trying to think of reasons why. Perhaps it's because the hens become a permanent part of your world and therefore deserve permanent accommodation?

Unless you are actively moving your mobile coop as part of a crop rotation process or similar, most mobile coops find their natural spot and end up staying there. Also, a well-designed fixed coop is easier to clean and manage than a mobile coop, so you can see why people end up with fixed coops.

Now, what do you want to know about fixed coops? If I were you, I'd want to know how to make a good one, so that I didn't have to make a few before I got it right.

Most importantly, make it easy to clean. A little thought and effort here will pay dividends for years to come. So, a sealed floor, concrete or bricks or pavers, anything smooth that will make it easy to sweep out (as well as stopping Mr Fox digging his way in). On the subject of flooring: the idea is to keep water out and keep it dry. If you do this, you can line the floor with some disposable covering, making for a quick and easy cleanout each time. Shavings, newspaper – preferably something biodegradable that can be composted or taken as part of a green waste collection, avoiding landfill.

Another element to a good fixed coop is headroom. It is so much easier to clean or maintain a coop if you can stand up inside while you're doing it: well worth a little extra expense in building materials. Having removable perches will make the sweeping out task easier. Instead of screwing the perches into the walls, make a simple u-shaped bracket for them to drop into.

If you make the nesting boxes accessible from outside the coop, you won't have to go inside to get your eggs.

I have mentioned overheating before, so please note: when you look into coop designs and pictures, you'll see many with enclosed roosting areas. A lot of these pictures come from North America where it snows in winter. If that isn't your climate then these are not the styles of coops for you.

If it's warmer where you live, envisage an aviary. You want at least one wall of mesh. If you think of that as the front of the coop, then you can have mesh halfway across each side as well. The other half of each side and the back should be solid to provide some draught protection for the perches.

Mobile coops

Mobile coops are great. They are perfect for first-time chicken-keepers and come in many shapes and sizes. The big ones used on free-range farms are towed around on skids by a tractor: not relevant to the backyard chicken-keeper, I know, but I think they're cool. The big advantage of a mobile coop is that while you experiment with how your hens are going to fit into your garden, you can try different locations. Another benefit is that the spot you might want to have your chickens in summer (in a shady hollow for example) might be a very poor location for them in winter (because it gets boggy, for instance).

I supply A-frame coops with wheels, which makes moving them around pretty easy. I have had customers who used their coop like a lawnmower, moving it daily or every few days to slowly cover the lawn. The hens 'mow' the grass under the coop in the morning before they are let out, and the coop is moved frequently enough to prevent the hens denuding the grass under it. This works for some people, but the effort seems a little too great for a lazy chicken-keeper like me: each time you move the coop, you end up raking up fallen straw and hosing the poo into the grass.

Now, just because you *can* move the coop, doesn't mean you *have* to. I have many customers who find just the spot for their coop and it stays there. I get asked all the time whether the coop needs to be on the lawn. The answer is nope. A mobile coop can go anywhere you like, with the proviso that you will need to clean under it eventually. If it is on dirt, lifting it off and raking the excess poo away is quite easy, but perhaps not so easy on pavers or concrete. Just make sure it works for you. The less time you spend cleaning up after your hens the more you will enjoy them.

My final word on mobile coops is that they still need to be fox proof. To my mind that means they need a mesh floor. Now, the mesh floor doesn't need to be integral to the coop: it can be separate and the coop sit on it. I have even seen coops with flaps of mesh that drop down from the sides.

If the coop is sitting on separate mesh, there needs to be some mechanism to stop the two becoming separated and providing a point of entry for a predator. If the mobile coop you fancy doesn't come with a mesh floor (some will try to tell you that it's not necessary), it is easy enough to buy some builder's reinforcing mesh or fencing mesh separately and put the coop on that.

Previous page: A flock of hens will save you time and money on mowing overgrown grass. Opposite: An A-frame mobile coop with feeder and waterer made from recycled plastic bottles.

The chicken house

I recommend a reasonably heavy grade and galvanised. The heavy grade means it is tough enough to pick up and shake or sweep for cleaning, and the galvanising stops it rusting (it is continuously damp under a coop).

Space requirements

How much space will your hens need inside the coop? The actual minimum space required inside the coop is remarkably small, but it is dictated somewhat by climate. In a cool climate that is never warm overnight, the hens won't need as much space because you don't have to be concerned about overheating. The barest minimum in this situation is enough room for them to perch and get in and out.

Rather than give measurements for length of perching space, it is easier to think in terms of chicken widths. If overheating *can* occur, I would say best to leave a minimum of two hens' widths for each bird, to allow them to separate when it is hot – if needed, you could reduce this to 1.5 hens' widths. Of course this varies with the size of your hens, but don't be stingy unless you have to. If you make your coop too small it will be a chore to clean and maintain. Small and, more importantly, overcrowded coops can contribute to chicken stress and make them more prone to disease. So, if you can afford it (in terms of garden space and financially) go large!

If your hens are to remain locked in the coop when you go away for the weekend, you will need some ground space within the coop, and the more the merrier. With larger coops, this ground space becomes, in effect, an integrated hen run.

The smallest coop I make, designed for two hens, has a floor space within the coop of a little over 1.5 square metres (16 square feet) and I wouldn't provide less than that if your hens are to spend any daytime in the coop.

Interestingly, this amount of space exceeds the definition of 'free range' proposed by the Australian Egg Corporation, which wanted to apply a stocking density of 20,000 hens per hectare. That's two hens per square metre! You can imagine what I think of that definition: it's (*insert expletive here*). The organic stocking density for free range sounds more appropriate at 2500 hens per hectare if the range is rotated and 1500 if the range is fixed. For the UK (at least while it is in the EU) the definition of free range limits

Opposite: 'Go away, I'm busy.' The perfect nest is warm, dry and dark with plenty of nesting material.

stocking density to 2500 hens per hectare unless you meet some rotation exemptions. In the US, free range is a free-for-all – all a producer has to do is demonstrate hens have been allowed access to the outside. No stocking limits at all. Bonkers!

If you are forced to buy eggs, some companies are now putting the stocking density on the carton. Remember, 10,000 hens per hectare is one hen per square metre (11 square feet) and given that they aggregate, especially around food and water, that doesn't sound very 'free range' to me. More like a barn with a grass floor or free range in a Tokyo subway during peak hour. Better than cages, without a doubt, but can't we afford better?

Building your own coop

Give it a go: no matter how rubbish you are on the tools, the outcome will still be an achievement to be proud of. To build a chicken coop you really only need a cordless drill/screwdriver, drills, a hammer, a saw and some pliers. As long as you meet the minimum requirements outlined in this chapter, your coop will be fine. It might not be a thing of beauty though.

Pointers to make the job easier:

- Hardwoods are harder to work with, but last longer. If you want to put a nail or a screw into hardwood, especially the tough-as-old-boots Australian hardwoods, best to drill a hole first or you will most likely snap the screw or split the timber.
- A sheet of plywood makes an excellent workbench that can be placed on anything stable. If you are using trestles, screw through the ply into the trestles to make your bench more rigid. Because plywood has right angle corners, you can use it to lay your work on and it will be pretty square. You can also draw on it to lay out your plans, and an eraser will remove the pencil marks. A 12 mm (½ inch) thick sheet is fine; 15 mm or 18 mm (¾ inch) is more expensive, but stiffer.
- Timber that is in direct contact with the ground doesn't last well. Most timbers rot if they stay wet for any time, although hardwoods last longer. If the coop is fixed, you might have to consider termites. A mobile coop is safe from termites as long as you move it around or at least lift it up occasionally and clean it.

Y Y Don't be tempted to use copper chrome arsenate (CCA) treated timber: it is not suitable for a food generating environment. There are other less nasty timber treatments, but I'm still not keen on treated timber in a coop, especially in or around the nest.

Converting an existing structure, such as a shed, cubby house or aviary

With just a few hours' work an old tin shed can be turned into an excellent chicken coop. Sheds tend to be sitting on concrete slabs as well, and are often available pretty cheaply second-hand. Use an angle grinder to cut out panels in the side for ventilation. (Marking the areas to cut out with a permanent marker, rather than just blazing away, will give you a neater job.) A few minutes with sandpaper to clean up the edges is worth the effort; wear gardening gloves if you are worried about cutting your hands. Then you need to fill in your new ventilation panels with mesh. The easiest method is to use the discarded panels: cut a 25 mm (1 inch) wide strip off each side of the discarded panel, and use these to sandwich the mesh between the shed wall and the strips. Pop rivets are better than screws for attaching these so that they don't leave sharp screw tips on the inside.

Screw in some perches, find something suitable for a nest box and you have a coop all done in time for beer-o-clock.

If you don't have a slab (because you bought an old shed, or decided to relocate your shed), you can lay your shed on big concrete pavers or pour your own slab. Don't panic if you have no idea how to do it: just spend half an hour watching a few videos on the Internet to see how different people approach it and then dive right in.

A cubby house no longer frequented by progeny is another great option for a coop. Add ventilation, perches, a nest box of some sort and you are sorted. Check that the door(s) close securely and have a think about how you are going to clean it out. But if your cubby house is an old treated-pine one, demolish it – it's most likely to be CCA timber and reusing it is a false economy if there's any risk of arsenic poisoning.

Aviaries are pretty well ready to go – just add the perches and nests.

Overleaf: A coop within a coop. This can happen when you fall in love with chicken-keeping and expand your flock.

The chicken house

Buying a commercially made coop

When I started Rentachook in 2001, there really weren't any companies making chicken coops for the backyard chicken keeper. So I had to design and make my own. My, how things have changed since then. There are a number of aviary makers these days who make coops too, and there are many little cottage manufacturers of mobile coops. And there is a wealth of mass-produced timber coops made in China.

Being a manufacturer of chicken coops myself, I have to be quite careful about what I say here. I will start by saying that every product involves compromise.

There are beautifully made chicken coops made of materials that will last forever, but quality materials are expensive, and quality workmanship takes time (and, therefore, money), so you will pay for these. Unless you are minted, these coops make a good upgrade rather than initial purchase. I have never regretted paying properly for a quality product that is well designed and fit for purpose, but I'm just going to throw in the thought that, as a beginner, you might want to make sure chickens are for you before spending the big bucks. On the other hand, you will always find a buyer for a quality product if you want to sell second-hand. Another benefit of buying quality is that the vendor is usually more willing to provide pre-purchase information and post-purchase advice.

At the other end of the scale, there are mass-produced timber jobs, imported flat-packed. I have no problems with anything being made overseas, as long as you have good quality control. I suspect these coops are primarily made for the European and North American market, because they appear to be designed to protect hens from the effects of cold rather than heat, which is usually a bigger concern for the Aussie or Kiwi chicken keeper. These coops look like your mental image of a coop and are reasonably suited to the task. Their outstanding feature is that they are cheap. Well, they are if you buy them online, rather than at the local hardware store or a pop-up 'specialist' coop website.

Let your mouse do the walking if you're thinking of buying one of these, as the same coops are sold all over at vastly varying price points. Oh, and avoid a schoolboy error by checking the published dimensions with a tape measure. This will ensure you don't buy a guinea pig hutch by mistake.

These are not bad coops for a beginner, but don't expect them to last. They can be modified to improve their suitability for warm weather – knock out internal partitions that enclose the roost, or make another roost out in the open. The upside is price. The downside is durability: doors can jam frustratingly when the wood swells in the wet. Above all, make sure it is fox proof: it needs a mesh floor if it is not set on a sealed surface and they usually don't often offer this option.

Hen runs and fencing

If you are planning to let your hens roam free with the run of the whole garden all the time, skip this bit. Go and read the paper, find out what's going on in the world...

A hen run does make life a bit easier as it means that you can let the chooks out at your leisure and return them to their place if they start to take over the garden. Now, the art of a nice hen run is to have it blend in with the garden and not look like a henitentiary. You know exactly what I'm describing: when you overdo it and your place looks like Alcatraz, only without the water views.

Many people who get into chickens also get into gardening and food growing. For you, the choice is either: contain the hens or contain the vegies. If not, say *sayonara* to the soft-leafed yumminess in your vegetable patch. The chickens love it every bit as much as you do.

A 900 mm (3 feet) – waist-high for most of us – fence of chicken wire is sufficient to keep most hens in or out, if their wings are clipped (see page 139). You can go to 1200 mm (4 feet) and upgrade the wire to aviary mesh but, honestly, you won't need to, unless you have a Henny Houdini. I also think it just makes too big a statement about confinement to have a fence that imposing.

To hold up your chicken wire fence, tomato stakes will do well enough but 50 mm x 50 mm (2 inch) hardwood stakes look better. Chicken wire can be attached to wooden stakes, or a paling fence or whatever, by screwing screws halfway into the timber and wrapping the wire around the screws a couple if times. Cable ties work well too but make sure you clip off the ends of the ties. Nothing looks quite as half-arsed as cable ties with the loose tails sticking out in the wind.

Henny Houdinis do exist. There are truly some hens that are

extraordinarily good at escapology. Luckily, such skills are the preserve of the few, not the many, and there are some tricks to foil their efforts.

Hens with properly clipped wings can't fly predictably up and over a fence. However, they can hop from vantage point to vantage point if they are so inclined and this is the route for most escapes. In the same way that you should not place garden furniture up against pool fences, you should check that you haven't unwittingly left a ladder with conveniently stepped stages that allows a hop over the fence.

Hens find it easier to get over a fence if the top is rigid and can be used as a final step. Floppy-topped fences are both easier to build and harder for hens to get over. To make escape even harder you can string a single strand of wire 100 mm (4 inches) or so above the top of the fence. The hens aim to use the fence as a springboard to freedom, but hit the wire and fall back into the hen run. Curses, foiled again! This trick works really well, looks less intrusive than a fence of that height and can also be used on top of existing low fencing or retaining walls.

To get into your hen run, make or buy a simple framed gate. It is worth it. Yes, you can climb over the fence if you are tall enough, but if it's muddy and you slip, you will take the whole fence down with you. I choose not to confirm whether or not this has happened to me. If it has, it might or might not have happened in the morning when I was dressed for the office.

Mulch beds and deep litter

Speaking of muddy, if you choose a damp part of the garden for your hen run, it could well become a quagmire of biblical proportions. It is fine to be shady, but it shouldn't be boggy: something that gets wet but drains well is ok.

My preferred method of managing potential bogginess in the hen run is to have a deep litter mulch. The great advantage of this is that within the mulch there are plenty of grubs and bugs that the hens love to eat; they turn it over continuously looking for them and, in doing so, they bury their poo. If done right, this is largely self-managing. Pretty much any woody tree mulch will do, the woodier the better. Too high a proportion of leafy stuff and it doesn't last and composts down too quickly. My preference is woody eucalyptus mulch, but you don't have to be too specific.

Opposite: When the hen run is bare, throw in grass clippings from the lawn-mowing as a treat. Note the floppy-topped fence here to stop feathered escapologists.

Woody is better, and hardwood is best if you can get it. Lay a covering of 100 mm (4 inches) or so over the whole hen run.

Some councils offer free mulch to residents so this is worth checking out. Failing that, it's cheaper to purchase from tree loppers/chippers directly, rather than from a landscape supplier. You will get the best rate by taking the whole truckload and being willing to wait until they are either in the area or have too much mulch to deal with.

I find that a good mulching lasts about a year before it becomes basically soil. But what good soil it is! Before adding a new layer of mulch, I always mine some of this soil to use elsewhere in the garden. The woody bits can be screened out by leaning some stiff mesh against a wall and throwing the composted mulch at it. If you get your angle right, the soil falls through and the woody bits trickle down the mesh and collect on the outside.

Cleaning out the coop

There is definitely a sweet spot in terms of the frequency of cleaning out a coop. Too frequently, the task is easier but you do it so often that the time adds up. Too infrequently, and the extra work involved negates the time saved.

For a fixed coop, my preference for floor covering is dust-extracted wood shavings. Shavings come in bales as bedding material for all sorts of critters, and they are usually very dry and so desiccate the poo that lands on the floor. Over time, your poo-to-shavings ratio gets too great and then you just sweep it out. If you leave it too long and it stays moist, the task is unpleasant. You can use a tray under the perches to extend the lifespan of the floor shavings. The tray should be covered with mesh so that the poo falls through but the hens can still walk on top – this prevents the hens walking in their excreta.

There are techniques for using deep litter as coop flooring that are supposed to be largely self-managing. I don't have personal experience of these to draw on, but they can go wrong and, when they do, you will have a lot of material to deal with. Deep-litter systems are better suited to the hen run, in my opinion.

I'm not big on hosing coops out as a regular method for cleaning: you just end up dealing with a semi-liquid mixture of poo and bedding material. Better to sweep it out dry and wash it separately, if washing is actually required. Once the coop is swept out, I throw down some powdered hydrated lime before replacing the shavings. The lime is a natural disinfectant and makes the ground under the shavings an inhospitable residence for insects like mites. Wear a dust mask when you're doing this or you'll give your lungs a good cleaning too. It is cheaper to buy builders' lime in 20 kg (45 lb) bags from a trade hardware store, than gardeners' lime from a nursery. (Spend the savings on organic feed.) Powdered sulphur works well for this too, but is a bit more expensive than lime.

There have been occasions when I have had to fully disinfect a coop. It is not easy to do. No, scratch that, it is a proper pain to do properly and unnecessary in most cases. I have, on occasion, purchased batches of hens that have had transferable ailments and I had to ensure the condition didn't transfer to the next batch of hens. For you, don't bother: it's a chicken coop, not a guest bathroom. You would only need to consider full disinfection if you had an outbreak of disease or parasites as described on page 194.

It is not a bad idea to give your coop a thorough clean once a year. A wash down with hot water, or high-pressure water, is better than cold, and high-pressure hot water is the best. Wait for dry weather so the coop can dry out; if you are keen you can spray some disinfectant around as well. Disinfectants suitable for poultry are available from pet stores and stockfeed or farm supplies stores.

For mobile coops, you want to deal with the build-up of chicken poo, especially under the perches. This is easy if the coop is on grass and moved regularly: just hose it in. If the coop is on dirt or is stationary, lift the coop off and stand it up so you can give the bottom a good brush down with an outdoor broom. Rake out the excess poo from the ground and throw down a handful or two of lime for good measure.

Poo, soiled shavings, used nesting straw and other such materials go well in the compost and, as long as you're doing your compost properly, the heat and biodegradation processes will kill pathogens and other nasties. Discard such material if you have infection or parasite infestation: you don't want any possibility of reinfection. If you live in an area with garden waste collections, poo and bedding material can usually go in your 'green bin'.

Nesting materials

Now we get to the vexed question of suitable nesting material. Do you have to use straw? What about shredded paper? Sugar cane mulch? Pea straw? Wood shavings? Argh! Too many choices!

In truth, any of the above will do. In autumn you can also use dry-fall leaves. Rather than tell you which is best, I will outline the benefits and deficiencies of each:

- There's nothing that says 'nest' so much as golden (wheaten) straw. A clutch of brown eggs nestled in a shaped dish of straw is a human's picture-perfect image of what a nest should be. Straw is an excellent nest material because it's resistant to humidity and durable. The downside is that a bale of straw can be surprisingly hard to get if you live in the inner city. I've found it isn't routinely stocked at the big hardware stores and nurseries.

- Close to straw is pea straw or sugarcane mulch, both of which are readily available at hardware stores/nurseries and have the advantage of being bagged, so you don't need to spend ages picking straw out of the carpet in your car boot. They are a little finer than straw but still work well.

- Shredded paper has the advantage of being free, but you would have to be a pretty dedicated Parsimonious Poultry Person to baulk at the cost of a bale of straw or cane mulch every few months. A factor to consider with shredded paper is the porosity of eggs. Eggs have about 6000 little air holes and can wick up chemicals, including from the inks on printed paper. This might sound fussy, but a number of my customers have been organic super-purists and brought this to my attention. So, I thought I'd share it. Paper is also more hydroscopic than other litter materials, meaning you have to change it more often and it can leave unappetising ink stains on the eggs.

- Another alternative is wood shavings. I use wood shavings for the floor of my coop, but they are also eminently suitable for the nesting area and have one big advantage: if you have one of those super-annoying hens that likes to roost in the nest, the dry shavings suck the moisture out of the poo and less of it gets onto the eggs.

107

Opposite, nesting materials, clockwise from top left: Coarse wood shavings; Pea straw; Lucerne straw; Fine wood shavings.

The chicken house

Feeders and waterers

The most commonly available feeders on the market are the hanging style, where you pour the feed in the middle and the hens eat out of a ring around the edge. These must, in my opinion, have been designed by stockfeed companies with an eye on increased profits. A global conspiracy to increase stockfeed sales might be a little far-fetched, but these feeders are terrible for wasting feed. I have never seen one that didn't have a ring of spilt feed below it.

At the other end of the scale, you can buy feeders that remain closed until operated by a hen to open them. They certainly prevent vermin getting a meal after dark, but you pay for what you get and I've always been too tight-fisted to buy one.

There are many different types of feeders on the market: too many to go into the relative merits of each. Some boast the benefits of storing a quantity of food, meaning you only have to top them up occasionally. For me, I have previously argued that you need to be interacting with your hens daily if for no other reason than to collect the eggs. So, how hard is it to top up their feed each day?

My suggestion for the Parsimonious Poultry Person is to make your own feeders out of leftover plastic milk and juice containers. Just cut a hole about the size of an apple in the side opposite the handle about halfway up. I find kitchen shears are excellent for this: just stab the bottle in the centre of the cut-out to get you started. To stop the hens knocking the feeder over, make a hook from an old coat hanger or similar wire and attach it to the handle. Don't worry if you mess the first attempt up: the raw material is in plentiful supply.

Here's a trick: don't fill a bottle feeder, or any other top-up feeder for that matter, right to the lip of the opening. If you do, the hens will pull the feed out and spill it on the ground. Fill it to a couple of centimetres below the lip to minimise hen spillage. Some hens are quite determined to pull feed out of the feeder – if you have one of these, make bottle feeders with increasingly smaller holes until you get one that solves it. And overcome your fear that the hen won't be able to get her head in and get a meal: hens have quite prehensile necks.

Whatever feeder design you prefer, you need to make sure the feed doesn't go stale because the hens cannot reach the bottom. If you're using a top-up type of feeder, periodically pour out the

Opposite: A White Sussex using a drinker that has been easily modified to prevent the birds using it as a perch.

residue: once a week in wet weather and once a fortnight in dry weather should do it. You don't want some of the feed to stay there indefinitely and turn manky.

As for waterers, otherwise known as drinkers, there are plenty of options available. Plastic bottle feeders are perfectly sufficient as waterers for you parsimonious types and if you leave the lid(s) off, you can top them up with a watering can without having to unhook it or even bend down. (I'm all for the lazy option.)

You can also use fancy self-replenishing gizmos if you lean toward the gearfreak end of the spectrum. Whatever you use, the waterer should be placed in the shade in hot weather, as warm water is less refreshing than cool. Equally, if there is the potential for freezing, you need to check the waterer hasn't iced over in the morning and break into it if it has.

The most important thing is that the hens don't run out of water in hot weather. A hot summer day can kill a hen without a drink. I recommend two separate sources of water in summer and don't be afraid to add some ice cubes.

Water should be changed often enough that it doesn't turn green, and the waterers cleaned periodically. If the waterer looks or, worse, smells rank, it is time to give it a scrub. A dash of apple cider vinegar or squeeze of lemon in the water can reduce the frequency of cleaning.

Opposite, clockwise from top left: Do not overfill the feeder or the hens will spill it; Refilling a hanging waterer; A simple home-made waterer; A hen-operated feeder is more expensive but encloses the feed when not in use.

The chicken house

6

FLOCK MANAGEMENT

Catching and handling chickens

When you own chickens, at some point you will probably need to catch one. This can be surprisingly hard to do, especially if you have a bantam that can accelerate like a Formula 1 and turn on a dime.

A chicken will continue to run until it can run no more. If you have the stamina, it is possible to chase a chicken unto complete exhaustion, whereupon it will stop and keel over and you can just lean down and pick up the immobile bird. I've never done this (I'm nowhere near that fit) but I have had it described to me as quite dramatic.

As an alternative you might choose a less physical approach. Consider, if you don't actually have to catch and hold the hen, it is easier on both of you to just herd her if you want her out of your vegie patch. Herding is easier with a long stick such as a broom handle or tomato stake in either hand, giving you a bit of directional control. You can also use this arm extension to give the slowcoaches a tap on the tail feathers.

If you are catching the hen to, say, administer medication such as a worming tablet or to clip a wing (page 139), why not save your breath and do it in the evening? After dark the hens are much

Opposite: There are many ways to hold a chicken (this one's a Speckled Sussex) but it's hard to go too far wrong.

slower moving and dopey. Often you can just pick them off the perch, do what you've got to do, then plop them back down and move on to the next bird. Kind of like a hen vending machine.

Chook-hooks and nets

For daytime catching, good technique or implements can make the task easier. Starting with implements, a chook-hook works well: think a diminutive version of a shepherd's crook, but for use on the legs not the neck. You can make one from heavy-gauge wire or light metal rod; the bight of the hook should be a bit wider than the lower leg (the scaly bit). Keep this width for about an inch to make a channel to catch the leg in and then splay the end of the wire out to make catching easier. Once you get a leg in the channel you pull back and catch the hen by the ankle above the splayed toes. With a practised hand it is a quick push-pull motion and the hen is yours.

Another implement is a fishing landing net. Easy there, Tiger, before you cry 'cruelty', think: is it crueller to catch a hen quickly in a net or with a hook, or stress it out with a protracted chase around the garden? It's your call, I'm just providing options. But, on the whole, implements are more for the poultry professional who has to catch hens regularly than for the backyard hen-keeper.

117

Other catching options

The tamer the hens, the easier they are to catch and crossbred layers are often more easy-going and amenable than purebreds.

My preferred hen catching technique is something I learnt from a huge goanna during a camping trip (for the uninitiated, a goanna is like Smaug without wings and this one was 2 metres/ 6 feet long). He approached slowly and steadily, but then suddenly upped the speed to a charge. The change of pace got me scared and I have to admit that I bolted, leaving him free to eat my breakfast. This also works well on chickens. A sudden change of pace often catches them unawares and even if it doesn't, they sometimes go into the submissive pose I call the 'stomp-dance'. You will know this when you see it.

Another thing to know is that you don't have to grab a hen to catch it: a flat hand on the back and quite gentle downward pressure is often enough to immobilise a hen, allowing you to scoop it up.

Opposite: A broom handle or stick makes an excellent arm extension to herd chickens when catching is not necessary.

Flock management

How to hold your hen

As for handling hens, they are not unlike rugby balls with legs. Pick them up as you would a rugby ball, two hands over the wings holding them in to the body, and fingers underneath. You can also tuck them under an arm as you would a ball or lay them along an arm, holding the legs.

They usually calm down quite quickly, and more quickly if you contain and minimise the flapping. I have been known to transport particularly tame hens on my shoulder. I don't do it for long because chicken poo down the back is a little antisocial.

(Holding the hen by its feet and body hanging down won't cause harm, but is not very nice. It is, however, expedient for the professional as it allows holding and transport of up to 10 hens.)

Everyday chicken behaviour

Without a doubt, chickens are crazy critters and there are some chook behaviours that defy categorisation.

Dust bathing is the first one that comes to mind. Hens wash themselves by fluffing dirt through their feathers. They lie on their side and flap dirt through their plumage, often with their head lying to one side. The first time I saw this I thought the bird was in the final death throes of mad-chicken disease.

So, ideally, somewhere within their range there will be an area of dry dirt that the hens can use for dust bathing. Usually, it will be under the canopy of a tree where the soil stays dry and dusty, but it can be anywhere. It is important that your hens have access to such a place as it reduces the risk of them getting mites or lice.

Another strange behaviour is sunbathing. Sometimes even without the sun. So don't freak out if you see one of your hens lying stupefied on one side in some random place.

I also want to pre-warn you about something I call 'brain-freak'. The hens seem to be acting naturally (that in itself is an odd concept for chickens) and then, for no obvious reason, one just snaps and bolts helter-skelter across the yard as though being pursued by one of the Horsemen of the Apocalypse. Just as suddenly, she stops and returns to normal as though nothing out of the ordinary has happened. I can't explain it, but if you see it, put it down to the madness of nature.

Opposite: Hold your hen like a rugby ball. (No roll passing, though!) Overleaf: Hens 'wash' themselves by fluffing dust through their feathers.

Raising chicks

I counsel that it is easier for your first hens to be 'point of lay' but there is an undeniable irresistibility to the cuteness of chicks. Since raising chicks is undoubtedly a wonderful experience, I also advise you give it a go at some point in your poultry journey. There are many ways to go about raising chicks, the easiest being to place sexed one-day-old chicks under a broody hen and let her do all the work for you. I'm not going to talk about incubators here as, frankly, I've never used them myself. I prefer the *au naturel* approach. So, read on to learn about broody hen and some other approaches.

Broody hen

For those unfamiliar with the terminology, a broody hen is a hen that has decided to reproduce and sits (or sets) on the nest day and night to incubate the eggs. Left to her own devices, she will wait until a clutch (batch) of eggs is laid and then sit down on them to keep them at incubation temperature and humidity for 21 days until they hatch.

　　To make this work you will need either a rooster or fertile eggs. If you do have a rooster, he will know his business. (I once had a small rooster that was bullied by the hens and really pushed around by them. He still managed to do his duty, although he had to time it right or cop it from the bystanders. I was properly impressed by his dedication. I could only keep him for a short while as no effort I made could keep his crowing at an acceptable level, and that includes boxing him up each night and hiding him indoors to prevent early morning vocalisation. I was so fond of him though that I arranged a country retirement. He ended up at the top of the pecking order in a flock of 10 hens with no competitors for their attention. I understand he died happy, of exhaustion...)

　　Assuming you can't keep a rooster, fertile eggs are pretty readily available to buy and can be swapped with the infertile ones your hen is sitting on. Believe it or not, it's best to leave fertile eggs (called setting eggs) for a few days between laying and commencing incubating. Up to a week is good, max two weeks, but don't put them in the fridge in the meantime: a cool shelf away from sunlight works well. Ask the setting-egg vendor when they were laid. But bear in mind that if you leave your broody hen too

Opposite: Solitary confinement – a broody cage provides light and air all around the hen to discourage nesting behaviour.

long before starting incubation, you run the risk of her getting bored and running off the job before the chicks hatch.

You can cram a dozen or so eggs under a setting hen if you're super-keen but, depending on your luck, you could end up with 12 chicks and at least some will be cockerels.

That brings me to the downside of starting with eggs – half the blighters are going to hatch into blokes. Steel yourself for some culling and, if Murphy's Law comes into play, you could be eating quite a bit of coq au vin in the months to come.

As a general rule, you are not going to know how many eggs will turn out to be fellas for quite a while. People ask me if I can sex chicks and I reply: 'Yes, of course: when they start crowing they are cockerels and when they lay eggs they are hens.' This is not strictly true. I once had a bird that chased the ladies around, crowed and laid eggs as well. It freaked me out. When I looked into it, I was told that approximately one chicken in 25,000 comes out as a true hermaphrodite.

As your home-hatched chicks grow, you might notice a couple of them squaring off against each other. Dollars to donuts, those two are going to start crowing in a few weeks' time.

A few other points on broody hens:

- The hen can leave the nest for 15 minutes or so without problems and she should slip down to the hen cafe every so often to have a quick bite and some liquid refreshment;
- Sitting hens are more prone to getting mites or lice than others in the flock, probably because they don't get to dust bathe. Check her a couple of times for mites and dust the nest with a mite powder. Pestene powder is pretty benign and can be used prophylactically if you want;
- Hens will incubate eggs of different breeds and even different species. A mate of mine once put goose eggs under a Silky bantam, which she duly hatched. Within a couple of weeks the chicks were already towering over mum and would bowl her over when she called them to feed. Heroically, she continued with the maternal duties of defending her chicks from others in the flock until long after the goslings were by far the biggest birds in the hen run. Nature is just mad sometimes.

A sneaky trick to ensure you get an all-girl outcome is to wait until quite late in the set and one night swap out the infertile eggs she is sitting on with newly hatched sexed chicks. *Et voila!* Instant family and no roosters. To make this work you need to be pretty sure the chicks were laid that day (or yesterday). If they are older they won't imprint on each other and form the bond you need to have sitting mummy raise the chicks. Grill your chick vendor on the date of hatching and definitely don't take any chicks that are starting to grow feathers at the wings: cute fluffballs only.

It is best that the broody mum be given her own space to raise the chicks for the first six weeks or so, free from interference from the rest of the flock. The chicks are best fed on a starter or starter/grower feed, which has a different composition to the layer feed you feed the rest of the flock. Starter feeds often include a coccidiostat – a medication that prevents the chicks from contracting coccidiosis, which kills them pretty quickly. They also have a higher proportion of protein (around 20%) than layer feeds and reduced calcium and are better suited to chicks.

Buying chicks

No sitting broody hen? You can still bring home chicks and rear them. While it is tempting to get chicks from a classroom hatching project, the pitfall is that some will undoubtedly be roosters. Better to weather the storm of children's pleading than to acquiesce and have a 'hen' that starts crowing at 5 am.

Also, there are vaccinations for hens that are given to sexed day-olds and, if your poultry purveyor hatches in sufficient numbers, having them vaccinated is a bonus. Hobby breeders and purebreed people don't always vaccinate because some of the vaccinations come in quantities for commercial operators, say 300 bird batches, and they're not cheap enough to waste.

You can even buy chicks mail order; no kidding. It's not even cruel. (Just don't go and Google 'mail order chicks' – the results might or might not be what you were expecting.) Newly hatched chicks don't need food or water for two to three days because they have leftover reserves from the yolk. If you do manage to locate a mail order chick business, order them for despatch on a Monday or Tuesday so there is no chance they will end up in a post office over the weekend and go beyond the three-day window. A reputable supplier will have this sorted though.

Brooders

A brooder is basically any form of enclosure that provides your new chicks with protection, warmth and other basic needs until they outgrow it. For the sake of clarity I'm going to differentiate between a *brooder* for little cheep-cheep fluffballs and a *grow pen* for the buzzard stage.

The simplest form of brooder is a cardboard box that is open at the top so everyone can look down and admire the cuteness. If the first box is too small, the chicks will outgrow it quickly. On the other hand, if it is too vast it will not only take up the whole kitchen table, but will also run the risk of draughts, which can kill chicks, especially in the first week.

Little chicks are most sensitive to temperature and temperature fluctuations. You'll need to provide a source of heat that is constant for at least the first three weeks, or for the first six weeks in cold weather. The simplest form of heater is an incandescent or halogen light bulb. On this occasion, don't try to save the planet with a compact fluorescent or LED light: it won't cut it. The bulb needs to give off heat and the energy efficient ones just don't.

The drawback with using a lightbulb is that the chicks are in continuous light. This is good for the first 48 hours but it's best not to have continuous lighting after that. If you are using a suspended light bulb for a heater, turn it off for at least 30 minutes every day. Do this during the day as it's warmer then, and cover the brooder to give the chicks some darkness and retain the heat.

If you use a light as a heater, the chicks will cheep all the time, which is very cute... until it isn't. They will be quieter overnight if it is dark. They are more expensive, but you can buy special bulbs that emit only heat and not light.

You can also put a bulb inside a terracotta pot with a Lampmaker light fitting available from hardware stores.

If you are using a lamp, having a dimmer switch will allow you to adjust the heat easily, but don't go wiring it in yourself. I expect your insurers would take a dim view if you informed them your house burned down because you wired your own brooder heater.

The Internet is flooded with different options for heating things, and anything that provides a continuous supply of moderate heat will do: it doesn't need to be specifically designed for chicken brooders. That said, if you are going to raise chicks

Previous page: The pure joy of keeping chickens. Opposite: A brooder and heater don't have to be elaborate; a box and lamp will do. These chicks are huddling and showing signs of feeling a bit cool.

more than once, it is probably worth splashing a bit of cash. You are looking for a heater to maintain the brooder at a continuous 32–35°C (90–95°F) at the start and let the temperature drop by around 3°C (around 5°F) a week for the next few weeks until the chicks are living in the ambient temperature of 21°C (70°F). In summer you might only need to heat them at night after the first week or two, and then not at all after three weeks. In winter, longer obviously, but not generally more than six weeks.

The advantage of having a central source of heat, such as a hanging lamp or pot, is that the behaviour of the chicks will tell you if they are warm enough or too warm. If they are huddled together close to the heat source, then it is too cold. If they are concentrated around the edges far from the heat source, you guessed it, too hot. If they are all at one side they might be avoiding a draught. Your aim is for them to be randomly dispersed in the brooder and moving about like kids in a schoolyard.

A few more points on brooders and managing little chicks:

- Chicks are a favourite snack for other pets, so find a way to keep moggie at bay. You might need to eschew the cardboard box in favour of something with rigid sides that you can put mesh or a lid over. If you do have a rigid lid, make sure that there is sufficient ventilation and that the brooder won't build up heat and cook the chicks.
- Give chicks their first drink of water at brooder temperature. If it's cold from the tap, warm it up a bit and make it tepid. Chicks suffer from cold shock. If the chicks are day-old and have never seen water, you might benefit from dipping their little beaks in it. You can prevent the chicks drowning in a water dish by putting washed pebbles in it. (And, incidentally, ducklings should be allowed to dip their full head in the drinker but not swim in it.) Drinkers should be rinsed daily and the water can be further purified by adding a little apple cider vinegar (1 teaspoon per litre) or a thin slice of lemon.
- Feeders and drinkers can be bought or made from plastic bottles. Making your own is fun, especially with kids and, because the chicks grow so quickly, it allows you to make new ones as they grow out of the old. I find plastic milk bottles work best and they are easy to cut with scissors. You want the chicks to be able to eat the food, but not play in it.

131

Opposite: Chicks stay very cute for a week or so, before they sprout feathers during their less attractive 'buzzard stage'.

Flock management

- Chicks are messy little blighters. I recommend lining the bottom of your brooder with paper towels for the first few days and then moving on to shredded paper, dust-extracted pine shavings, or kitty litter pellets, all of which are good and absorbent. You can stick with the old-school favourite of newspaper, which works just as well. Don't let the brooder get too messy, or the smell can take the shine off the experience and the chicks can get sick quite quickly if the brooder is constantly wet and boggy.
- Chicks can have a problem with poo sticking to the feathers around their little bums. If left untreated, this can prevent them defecating. If you see poo stuck to the chicks, catch them and wipe it off with damp paper towel.
- After three weeks, place a perch in the brooder and encourage the chicks to roost on it. At this stage, a perch is any elevated spot for them to hop up onto at night. If you have a cardboard box brooder, pierce two sides of a corner with a stick to make a simple perch.

Grow pens and the buzzard stage

Alas, the fluffball stage doesn't last very long and the chicks start sprouting feathers, first at the wings and then all over the place. They look quite terrible when half-fluff half-feathers: you never see chicks or ducklings at this stage in advertisements for toilet tissue, and for good reason. This is what I term the 'buzzard stage'.

If it is warm enough, even little chicks can be allowed outside time, but remember to keep them under close supervision. Cats, dogs, birds of prey, even rats will go for a chick left unattended on a lawn. As they grow, the variety of beady-eyed predators reduces and you can cut down the need for supervision.

Buzzard-stage chicks are more likely to overheat than chill. Keep an eye on their behaviour and look out for them holding their wings away from the body, or panting. Start turning the light/heater off during the day and then at night as well. Once they are fully fledged (all feathers, no fluff) you won't need to worry about cold, unless you live in a particularly chilly climate.

Feed buzzard-stage chicks the rest of the starter/grower feed you bought too much of, or grower feed. By about 12 weeks they can be migrated to layer feed or continue on the grower feed you bought too much of until 16–18 weeks. Taper off one feed and blend in another in increasing proportions rather than changing suddenly – they won't respond well to sudden changes of diet.

A grow pen is a halfway house between the brooder and the coop, but if the chicks are your only birds, then the coop is the grow pen. Just keep them locked in it to keep them safe until they are close to full-size hens. If you have other hens, unless you have a broody mum looking out for them, they shouldn't cohabit yet, or the other birds will give them curry.

The grow pen is a means of allowing them garden time, outside time, but keeping them safe. This means that if you are out in the garden with them, they don't need any pen, if your other hens are in a hen run. Failing that, a simple wire enclosure will suffice. If you're not out with them, a small coop, a spare coop, guinea pig hutch or whatever will do. As they approach full growth, you will want them out of the house (if they are still kept indoors) and a spare or simple coop/hutch is easiest for that.

Now comes the challenge of introducing your chicks to your existing flock.

The pecking order

You have heard of the concept of a pecking order, right? (*You know, where the CEO gets the corner office with the view and they put IT in the basement. 'No, no, we really value IT, it's just that there's nowhere else for you to go and the basement is closer to your cable-connecty-thingies.'*)

Well, unsurprisingly, the term comes from chickens, who have a very well established order from Top Hen down to Nerd Bird. As with a corporate environment, the most Senior Hen occupies the highest perch and the Nerd Bird the lowest – how cool is that? Same as at the office!

Chickens, however, have nothing to aspire to because, once the pecking order is set, it doesn't change unless the higher bird dies or is removed. The other advantage of being Top Hen is you get first dibs on the food, water and pretty much anything else you want. The only time I have seen this dynamic broken is when a broody hen is mothering her chicks – no one takes on Tuff Mummy and comes out on top.

When you introduce new birds to a flock there will always be a bit of argy bargy to decide where the new hens fit in the pecking order. I have watched this many times and it's exactly the same as two men squaring off at each other in the pub. They stand up and try to make themselves look as tall and strong as possible. They puff their chests and chins out. They strut around.

There is nothing wrong with this behaviour (in chickens, that is) and it usually sorts itself out in a matter of an hour to a day, depending on the size of your flock and assuming this interaction doesn't turn into bullying. Once it's stopped you have your pecking order re-established.

Bullying

The pecking order is enforced by the flock and if any bird breaks the rules she gets – wait for it – pecked. Sometimes people confuse this enforcement with bullying. A bit of a peck here or there isn't anything to worry about, but consistent harassment, especially if the loser bird is being cornered and/or if there is bleeding and wounds, warrants intervention.

Believe me, when it gets bad you will know. I have seen Nerd Birds with horrendous gaping wounds and, once the wound is opened and blood is visible, other birds will chime in too and you need to separate them quickly if you want Nerd Bird to survive. Don't panic though: I have also been amazed at the full recoveries made by badly wounded birds.

If you suspect you are seeing the early stages of bullying of this sort, look for thinning of feathers or missing feathers at the base of the tail. For some reason this is where the Nerd Bird usually cops it. You will also often catch Nerd Bird being hemmed in a corner by Top Hen or her cronies, head tucked away from her assailants, which might be why pecking often occurs at the base of the tail.

If bullying is suspected, check to see if your chicken yard has any or all of the hallmarks of a prison (henitentiary): too many inmates and too small an exercise yard; not enough nutritious food; no rehabilitation activities; or inmates kept in their cells all day? Bullying is much more likely if the hens are poorly fed, overcrowded or bored. I'm also told there are some breeds that are incompatible, like Silkies and Game Birds – as I've never kept Silkies and Game Birds at the same time, I cannot speak from experience. You can always ask the breeder if new varieties are likely to be incompatible with your current flock.

Opposite: Silver-laced and gold-laced Wyandottes passing the time of day.

If you do have a bullying problem, there are a few things you can try without the challenges of permanently separating the hens. Try doubling up the feeders and waterers and space them apart so birds are not competing for food. Separate Nerd Bird for a while or, if you suspect that a disparity in age is a contributing factor, until the young bird is fully mature. Watching each other through a dividing fence can help settle things. Like a schoolyard, bullying is most likely when a new kid is introduced to the playground; the bully is going to assert their authority by taking the newcomer behind the bike shed for a reminder of who is boss.

Introducing new birds into the flock

Whenever possible, add new hens (plural) to your existing flock. If you only add one then this one hen becomes the focus of all the attention. By adding more than one the dynamic is more fluid. That doesn't mean that you can't add one hen to an existing flock, just that it's easier if you add more than one.

Always add fully grown hens to an existing flock; a minimum of 16 weeks old. Think of it this way: if you add a second-grade kid to a schoolyard with a rowdy crowd of sixth graders, it's likely the young'un will be used for target practice. No, seriously, for some reason an established flock will often all turn on a new 'teenager' or young growing pullet and peck it to death.

Add the new hens in the daytime and into the hen run or garden so they have plenty of space to get away if they turn out to be losers in the pecking order. I have had many customers say they have read on the Internet that you should add new hens into the coop at night. And that you should always have an even number of hens (as though a chicken with a brain the size of a pea could count the number of her peers). Unfortunately, the Internet is full of people with opinions and no accountability and this is rot. The reason I don't recommend you add hens at night when, admittedly, the hens are dopey, is because they are confined to the coop then. So when they wake up at dawn and notice the newcomer(s), there is nothing to stop them penning her/them into a corner for some biffo.

If you add hens during the day you can, if necessary, intervene. Try to curb your natural instinct to dive in. Going back to the pub analogy, a bit of squaring off is fine but if someone picks up a pool cue, call the bouncer.

Debeaking

If you have to intervene in bullying, then it means debeaking. This is trimming the tip of the beak to the point where the tender, growing, alive bit inside the beak is exposed. Think of it like a fingernail. If you clip it too close to the quick it is tender and you wouldn't want to type with it for a while. Same goes for chickens, if you want to stop them from pecking each other. Debeaking is also used to stop hens pecking and eating their own eggs (but doesn't work if the eggs are thin-shelled and breaking of their own accord). In a commercial situation hens are almost always debeaked, because this allows the farmers to have higher stocking densities and they get higher ratios of egg-yield-to-food-weight.

In the early days of Rentachook I was willing to sell debeaked hens, but never again after I overheard a little girl ask: 'Mummy, what happened to that chicken's face?' Also, there is debeaking and debeaking. Heavy trimming and offset top and bottom beaks are unconscionable in my opinion. However, 'tipping' just the end of the beak a bit is not so bad and doesn't impede the bird feeding normally. In fact, some commercial crossbreeds will grow quite a pronounced hooked beak if completely untrimmed and a little trimming in this instance is ok.

For the bully hen, you want the tip of her beak to be tender so that she doesn't use it as a weapon. This will not prevent her feeding properly and the tender bit will seal over in time as with the growing of a fingernail. The gentlest way to debeak a hen is to use sandpaper. I use 180 grit and brush across the tip of the beak, holding the head in one hand, the bird underarm and sandpaper in the other. She won't like it, but you will really know when you've got to the tender bit, because she will suddenly struggle much more. When she does this, give it one more swipe with the sandpaper and let her go.

Another way is to use a Stanley knife. With the help of a friend, lay the chicken and her head on one side on a flat surface. Use the knife to trim the tip of the beak in a single cut straight down. I don't use scissors, although you can. (I will use scissors to trim or even up the tip of a beak that has splintered, but they are not as accurate as a razor-sharp knife used when the bird's head is pinned. Best to use small accurate scissors such as nail scissors.)

137

Other everyday problems in poultry paradise

Chickens escaping your garden and being found up the road with the neighbours? Hens eating their own eggs or insisting on roosting in the nest? Not everything always goes according to plan in poultry paradise, but most problems can be easily fixed.

Clipping wings to prevent flying

Wing clipping is not cruel and does not involve any real distress to the hens if done carefully. What you are trying to do is create uneven lift from flapping wings to stop the hens flying predictably and therefore keep them on the ground. Believe it or not, hens can fly quite well. Not like an eagle admittedly, but you would be impressed at how far a wild hen can fly.

The easiest way to clip the wing of a hen is to ask someone who has done it hundreds of times to do it for you. Ask the people you buy your hens from if they are willing. Failing that, it is not a complex task. Take a set of durable and sharp scissors and place the hen on a table of suitable height. Use your left hand (assuming you are right-handed) to gently grab the upper segment of the right wing, the segment closest to the body, and fan out the wing so that it is fully spread.

You will see two kinds of feathers there: what I call the body feathers that keep the hen warm, and the much longer and bigger flight feathers.

With the flight feathers fanned out, starting at the back of the wing and using the scissors in my right hand, I cut the flight feathers along where these feathers meet the body feathers. The feather stalky bit in the middle can be quite hard to cut through but don't worry, it is like hair and dead except at the base, involving no pain to the bird.

If you cut too close, it is like clipping fingernails too close and the bird will feel it. You will also see a drop of blood come out. Sometimes it is more than a drop of blood but, before wracking yourself with guilt, it is not a big deal and always heals up with no lasting effects.

If you don't cut close enough then you won't get the effect you are after, which is keeping the hen on the ground. Most chicken escapes involve hopping from one staging point to another and this requires an accuracy they don't have if their wings are uneven. For this reason, you don't want to clip both wings.

Opposite: Healthy hens are alert and engaged in exploring their surroundings.

Flock management

While the wing feathers will grow back completely, I find that hens don't often require re-clipping. They just get used to being ground dwelling. Being live animals rather than consumer electronics, however, this is no guarantee.

Moulting

Annual renewal of plumage usually occurs in late autumn. Like the winter sales, for some it is all encompassing, and for others, the sartorial variation is negligible. Same goes for hens: some will barely lose a feather, while others look as though they have been prepped for the supermarket freezer section.

Moulting is normally accompanied by a cessation of laying (there are only so many things a girl can prioritise at once). Actually, the metabolic demand of renewing feathers is quite sufficient to draw resources from egg production. The duration of egg hiatus varies with the severity of the moult. Some hens will lay through without skipping a day, others will hold off providing eggs until spring. Depending on the season they were hatched, some hens will not moult at all in their first autumn.

Egg eating

Egg eating: I hate it. Some chickens will eat their eggs, resulting in a desultory breakfast for you of just toast. And once it's started it's hard to stop.

Just watch what happens if you happen to drop an egg (because you were carrying too many) in the hen run. All the girls come running. Try not to set yourself up to fail. If you drop an egg, stomp it into the ground and bury it. Also, collect your eggs at least daily and, if egg eating starts, visit the nest and collect them as often as practical – three times a day if you are out in the garden. Learn from my mistakes. I have a bad habit of not collecting eggs daily and, with so many hens laying, the nest gets full to overcrowding and a sitting hen can step on and crack an egg. The broken egg then gets eaten, one or more of the hens gets a taste for it and starts pecking the good eggs, and I start pulling my hair out in frustration at my foolishness.

Thin-shelled eggs can also be a problem that contributes to breakage and therefore egg eating. See page 203 for more on preventing and managing thin-shelled eggs.

Debeaking can also be used to prevent birds pecking their

Opposite, wing clipping, clockwise from top left: Hold and fan out one wing; Trim the flight feathers where they join the body feathers; Clip one wing only; She will forgive you!

own eggs. I have found that hens are much more likely to start eating their own eggs when they are deficient in protein. So, the first thing to do is to check the protein content of the feed. If it is below 14 per cent this could be the root cause of your problem.

Roosting in the nest vs broody

Some chickens are just too dopey to work out what the nest is for and instead lay their eggs in odd places. People will advise you to buy dummy eggs to encourage the hens to lay in a nest (I suspect they might be purveyors of finely crafted faux eggs). If you really feel the need to encourage laying in a particular nest, a golf ball will do just as well.

As a general rule, if you build it they will come. Provide a suitable nest, and most hens will gravitate toward it. By suitable nest, I mean a sheltered, dry, semi-enclosed, dimly lit place with suitable nesting material.

Some hens will roost in the nest and, because hens do a large proportion of their defecation at night, this means poo in the nest and poo on your eggs. Eggs can be cleaned, but should not be washed (see page 213), so it's better to discourage this behaviour.

Roosting on the nest is not to be confused with being broody. Roosting on the nest is where the hen is out and about, hangin' with the flock during the day and returns to the nest at night to rest.

If you have a hen that is roosting on the nest, try picking her up and placing her on a perch. You might have to move the nest if that can be done or exclude your miscreant from access to the nest at dusk. Also, look at the edge of your nest and see if it makes an accidentally comfy perch. A few horseshoe nails nailed into the lip of the nest will make it uncomfortable and resolve the issue.

Broody is entirely different. Broody is where a hen decides to sit on the nest all day to incubate the eggs and provide you with cute fluffballs. Some breeds are more prone to broodiness than others. Silkies, for example, are quite prone to broodiness and make good broody mothers if you are raising chicks. Your crossbreed laying hen such as an ISA Brown very rarely goes broody – that would affect the number of eggs you get so the propensity for broodiness has been bred out of them.

Broody hens don't lay. They lay a clutch (no, not the handbag: a batch of eggs is called a 'clutch'). They lay a clutch and then sit on it with all their feathers fluffed up to keep them all warm.

142

When they have set and are intent on incubating the eggs, they stop laying.

To be frank, broody hens are a bit of a pain unless you want to use one to raise chicks for you (see page 123). A broody hen will defend her clutch with a peck of your hand. Admittedly, not a mortal wound, but if she's sitting on your breakfast, it can be annoying. An unfertilised egg kept warm by a broody hen won't stay fresh for long and if your other hens are laying in the nest the broody hen is sitting in, then you have a problem.

Broodiness is in part a response to a nesty environment (warm, dimly lit, nesting material etc). So, to stop a hen being broody, create an environment that is the antithesis of that. Make what is called a 'broody cage': a wire cage with a wire floor and no nesting material. These work best suspended above the ground so that there is light and airflow all around the hen. Also, I'm led to believe that the swaying motion helps to speed up the end of broodiness.

Use aviary mesh or reinforcing mesh to make the cage and make sure the mesh on the floor has holes small enough for the hen to walk on. Include a perch and a feeder and drinker, and make sure you suspend it in a reasonably well lit spot, but not in direct sun.

143

Two or three days in the broody cage should suffice.

An alternative to the broody cage is another medieval-sounding machination: the dunk method. Every time you see your hen on the nest, pick her up and dunk her in a bucket of water – just enough to wet her underside is sufficient. Hens don't like to go back on the nest wet. It'll take a few dunkings, but you'll get there in the end.

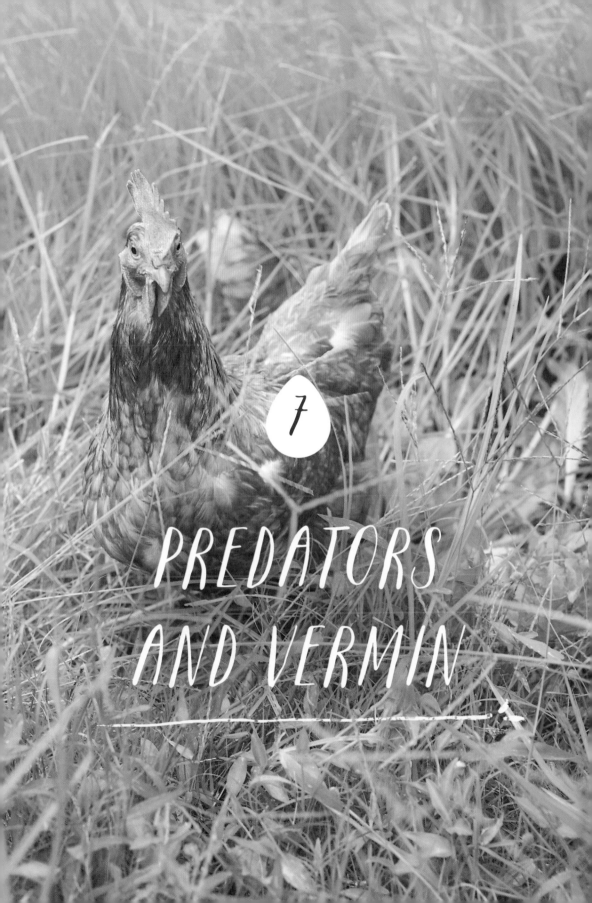

7

PREDATORS AND VERMIN

Predators

There's a reason we call a coward a 'chicken': chickens will run from predators. It's a good rule in life – run away from a fight with something nastier than yourself! Hens aren't high up on the nasty list, so as a chicken-keeper you have a duty of care to keep them safe. Don't worry: it's not hard to do.

Mr Fox

If you live in a country where there are foxes, Mr Fox is the number one enemy of your hens. Foxes are incredible predators and if you are a chicken they compare favourably to Ridley Scott's Alien. Foxes can climb trees, dig trenches and have earned all their accolades such as sly and cunning.

Once Mr Fox has located your hens within the area he roams, he will come and check on them intermittently, presumably in the hope that you will forget to lock them in. Mr Fox will keep coming back unless you find a way to discourage him.

So many times I have said to customers to be wary of foxes, only to be told there won't be foxes in their suburb. There is no schadenfreude whatsoever when I get a phone call weeks or months

Opposite: Silkies are gorgeous birds but, being small, they are more attactive to predators than bigger breeds.

Backyard Chickens

later requesting replacement chickens after Mr Fox has called.

The urban fox is almost better suited to the built environment than we are. All he has to do to escape any threat is jump over one fence. Think also that under every house that sits on piles and doesn't have a resident dog, there is a perfect, ready-made foxhole.

Foxes have been estimated to number more than 10 per square kilometre (100 hectares) in some suburban locales. And I have even seen foxes crossing roads in inner Sydney.

Mr Fox is usually a dead-of-night predator and his calling card is, most commonly, all of your hens with their heads bitten off, feathers everywhere, and one hen missing. There is no sugar coating: we are talking about a very distressing sight.

So, how to prevent fox attack? I have read that a 2-metre (6 foot) high fence, dug 30 cm (12 inches) into the ground, with a 50 cm (20 inch) outwardly angled extension will suffice to keep foxes out of your hen run. Also, that it can be made even more secure by adding one or more electrified strands.

Really, if you are going to do that, why not go the whole hog? Add a minefield and motion sensor controlled machine guns. How's that for a garden aesthetic?

You will never convince me that fencing will keep a fox out. I have also heard that male human urine will deter foxes. It might, but it might deter house-guests as well. I've got three sons who play in my garden... and I still lock my hens up at night.

Some say that having a dog will prevent fox attack and this argument has some merit. However, the only sure-fire way to prevent fox attack is to have a fox-proof coop and to shut the hens in each night without fail. There is more information on fox-proofing your coop on page 82.

I mentioned that Mr Fox is a dead-of-night menace and this is almost always the case. Bear in mind though that we are talking about a wild animal and so for this rule, there are exceptions. When food is particularly scarce, such as during drought in rural areas, desperation can drive foxes out at other times.

However, most fox attacks occur between midnight and dawn. I always recommend that people make it a night-time habit, similar to brushing your teeth before bed, to go out some time between dinnertime and bedtime to say goodnight to the ladies.

If you really can't bring yourself to make this effort, coops are available with automatic doors that open and close a hatch on a timer. Fancy options include solar-powered, battery-operated jobs, although I still advocate shutting the hens in yourself, for a calming pre-bed routine.

One final word on foxes. Mr Fox is often described as a malicious or indiscriminate killer for his propensity to kill all the hens yet only take one. Not so. Mr Fox kills all the hens to secure the food supply. If left undisturbed he will return and collect all of the hens one by one and take them to his larder. It is just that on most occasions, he never gets this opportunity.

Being hounded

Public enemy number two for the poultry keeper is the dog. I place these rankings based on the number of replacement hens I have sold to customers because of predator attacks. To give you an idea of proportion, I would say that for every 100 predator-based replacements, 90 would be for fox attacks, nine for dog attacks, and one for other or unknown (but unlikely to be fox/dog). I don't count dog attacks by people's own pets as predator attacks: that is more a pet management issue.

Usually the dog that kills your hens belongs to a neighbour, but if your property backs onto bushland or is otherwise not fully fenced it could be a roaming miscreant, bored and looking for some trouble.

A dog attack is most likely to occur in daytime and from an adjoining property. Bored dogs can move quite a bit of earth if they have a mind to. So if your neighbours have dogs, boundary fencing is important. Best to check it and that should give you the confidence to allow your hens to free range or at least be free in their hen run while you are out during the day.

I think that if someone's dog gets into your property and kills your pets then they should arrange, or at the very least pay, for their replacement. On many occasions I have had good, honest, yet somewhat guilty people call me to replace their neighbour's hens. Equally, if your hens were in the neighbour's yard, you can't blame their dog for your hens' demise. Wing clipping is your best solution here (see page 139).

One more comment on dog proofing. Please remember that chicken wire is chicken proof, not dog proof (or fox proof). A chicken-wire hen run is intended to keep chickens in, not dogs out (see page 101).

149

The horse with claws

This is how a feral cat was once described to me. I wanted to be sure the customer's hens were not killed by a house cat, as this would have been the first incidence that had come to my attention. He replied in as broad strine as I've ever heard: 'Nah mate, he's a monster; blimmin' horse with claws.'

I have since had only one other confirmed case of hens taken by a feral cat. In both cases the cat was observed attacking at night, in the same way as a fox. Therefore, if you follow the standard fox precautions, you should have nothing to fear.

Snakes

In all the time I have run my business in Australia I have never had a report of snakes being a problem with chickens. Not a single one. I think that despite having many of the top ten most venomous snakes in the world, we just have the right kind of snakes where chickens are concerned. In the US it is different: there you have a number of snakes attracted to the warmth of a coop or regularly available rodents.

Native carnivorous nammals

On mainland Australia there are dingoes in some areas and you will need to protect your hens in the same way you would against dog attack – with strong boundary fences.

In Tasmania there are no foxes; but there are Tasmanian devils and quolls. Both of these critters are protected so you can't do much except keep them out. They are both nocturnal, so use the same precautions as you would a fox.

The quoll has retractable claws and is an excellent climber. It also has quite a small head and can get through surprisingly small gaps, so coop security is important.

The Tassie devil is a powerful brute, stronger than a fox and with awesome bite power, so your coop needs to be pretty robust. Unlike foxes, Tassie devils will also hunt at dusk and dawn, so you might need to shut the hens in as soon as you get home and not let them out until a bit later in the morning. Quolls occur in some areas of mainland Australia too.

Previous page: A fully enclosed hen run means you don't have to lock up your hens every night. Opposite: Shutting the hens in for the night. Once they are used to the coop they will go in of their own accord at dusk – all you need do is close the door.

Predators in other parts of the world

In New Zealand there are no foxes and no native carnivorous mammals either. But it's not all good news: there are mustelids – ferrets, stoats and weasels, which were introduced to control other imported pests, rabbits and hares. Sounds a bit like the old lady who swallowed a fly to me. Defend your chickens from mustelids as you would defend from Tassie devils, as they are on the go at dusk and dawn. Mustelids, like quolls, can get in through small gaps – coop security again.

In the UK you need to be aware of foxes; your mustelids are native and protected and include my favourite animal in all the world, the sea otter. Another UK mustelid is the badger, which is enormously strong. For chickens, the badger is like a cross between a ferret and a Tassie devil with a bit of Rottweiler thrown in for good measure. You will need to build a strong coop to keep this bad boy out.

In the US there is an awesome array of creatures that will happily make a snack of your hens, including foxes, mustelids and coyotes, which are found mostly in rural, but sometimes urban, environments. Like Tassie devils, coyotes tend to hunt at dusk and dawn and take one chicken at a time, although they hunt in packs.

Raccoons will go for both chickens and eggs. They are omnivores and clever. Another downside is that they can carry diseases that can affect chickens and humans. I've read that they are mostly nocturnal but will feed during the day, and if they are a problem, you might need to build an enclosed hen run.

Opossums have been described as cat-sized rats. They are unlike Australian possums, which don't harm hens. They are nocturnal, so shutting the hens in at night should do the trick.

All raptors in the US are protected by law, so it is illegal to kill or harm them. Eagles need space to land and take off, so clever planting in the hen run can keep them at bay. Chicken hawks, including the red-tailed hawk, cooper hawk and sharp-shinned hawk, are efficient predators and dive at high speeds. Despite the name 'chicken hawks', their targeting of your poultry is more likely to be opportunistic than by design. There are also some owls that will take a hen, bringing us back to the necessity of shutting your chickens in at night. You can protect them from attack from above with netting or plants in the hen run.

Cottonmouth and copperhead vipers are among the species

of snakes that can cause problems for chicken keepers in the US. Hens will eat small snakes, so it is the big ones that are the problem. Snakes can be attracted by the warmth of your coop in cold areas, but might also come seeking rats and mice that frequent the coop for fallen food. Snakes can get through the tiniest of gaps, even through 2.5 cm (1 inch) square aviary mesh, although they might not be able to get out again with a full belly.

Vermin

I love the word 'vermin' – it's so evocative!

By vermin, I mean rats and mice, but the term can be just as easily applied to any critter that comes to eat spilt or leftover food. Possums, drop bears (the evil, carnivorous cousin of the koala), the lot.

Such creatures are ubiquitous. The art to coexisting with them is to ensure you don't supply them with a reliable food source. If you leave out spilt food, especially overnight, you are inviting the blighters to stay around and breed.

If you live near a river or a shopping centre, restaurant strip or whatever, rodentia may well already be there in numbers unlikely to be affected by the management or mismanagement of your hens. If this is the case, don't let anyone convince you that your hens are responsible for the plague of (two) mice recently observed by Mrs Tweedie, the neighbourhood busybody.

Some say the way to manage vermin is to vermin-proof your coop, and this can work – up to a point. But if you are managing feed poorly, mice and rats will find a way in somehow – it is just too tempting. You would be properly amazed by the size of a hole a mouse can get through. They can squeeze through anything they can get their heads through, and they have teeny heads when you flatten the fur down. For this reason, mouse mesh has holes only 5 mm (¼ inch) square.

Feeders are very important to vermin management, the priority being to have one that doesn't result in spilt feed. Some feeders are better than others and if the hens can pull the feed out of the feeder and leave it on the ground you are running the risk of a vermin issue.

It is not all about the feeder, though. What about kitchen scraps? With the exception of the UK (where it is deemed

illegal to feed kitchen scraps to your hens; see page 168 for more information on this), kitchen leftovers should make up an important part of your hens' diet (see page 166). They should be entirely consumed, or the excess removed, by nightfall.

Night-time is your problem time. Night-time is not your friend. Suffice it to say that if you leave food out at night, something undesirable will come to eat it. The easiest way I have found to minimise the likelihood of leftover kitchen scraps being a food source for vermin is to feed the kitchen scraps to the hens in the morning. I drop them on a dish in the hen run when I go down in the morning to let the hens out before work.

You really know you have a vermin problem when you see rats and mice during the day. This is because it is only out of desperation that a rodent will seek food during the day. I know it seems illogical that I have just told you that night-time food availability causes a problem that you see during the day, but bear with me. Rodents, like hens, have a social hierarchy. If there are too many of them, Nigel No-friends will not be permitted by the rat pack leadership to seek food at night and will be forced to feed during the day.

If you see rodents during the day, action will be required and this means baiting with poison or trapping. Whichever method you use, the main priority is to ensure your pets (and kids if you have them) don't get harmed. Once again, you can learn from my mistakes.

I used to use throw-packs of rodenticide that you can buy from the supermarket; the idea being you throw them into a spot that only a rodent can get to – which I did. I quickly learned that Border collie puppies are exceptionally adept at accessing seemingly inaccessible spaces. An after-hours emergency vet bill later, you would think I had learned my lesson. Not so. I went to greater efforts to locate the baits inaccessibly – and the dog went to greater efforts to get them. The next vet bill included cleaning the dog's teeth (and my wallet).

Feeling sheepish at my foolishness, I decided to go large and bought commercial bait stations, described as tamper-resistant. I wanted tamper-proof, but nobody wants to guarantee anything these days and, as it turned out, 'tamper resistant' was enough to keep a determined Rottweiler out. The bait station was chewed to unrecognisability, but the baits were unmolested.

Opposite: These feeders are popular but can encourage spillage, often leaving a 'halo' of feed around the base that attracts vermin.

Predators and vermin

So if you buy baits or set traps, you do need to make sure your pets don't get poisoned and your kids don't end up with broken fingers. Fortunately, there are a number of traps and products that enclose baits designed with the specific intention of keeping non-target species (such as children) out. You might end up having to purchase a commercial variant though.

Now all you have to do is ensure that your target species eats the bait or is lured to the trap. It sounds easy, but there are a few tricks that can improve efficacy. Rodents like to run along walls, so placing your verminator up against the wall increases your chances. Also, have a look at the corners of your coop for a point of entry; choosing a wall that intersects the suspected point of entry can help too.

If the baddies have a regular food source, remove it overnight and place the verminator of your choice where they are in the habit of getting food from. Vermin will also go for stored food, which must be stored in a lidded, reasonably robust container. If the container isn't reasonably robust a determined rodent can chew its way in. I use an 80-litre (20-gallon) drum which was used to import (quite a lot of) anchovies in its former life.

It's a good idea to keep straw in some sort of lidded container: stored straw makes an excellent home for a pesky rat.

Pigeons

I'm going to class pigeons and other feed-stealing wild birds as vermin too for the purposes of this chapter. While not in the same class as rats, mice and possums, they do come for the feed. Most of your disease problems and parasites in hens are brought in by these blighters.

Having wild birds come to your garden to nick food is part and parcel of keeping hens and, depending on the variety and number of birds, it can be somewhere between a pleasant addition and a royal pain. I have had pigeons so greedy that they ate so much that they couldn't walk, let alone fly.

You can minimise the number of visitors by keeping the feeders in the coop and out of sight. Managing or preventing spilled feed helps too. I have tried plastic owls to negligible effect and old CDs strung from branches so that they spin and flash with reasonable success. Best of all, hope for a wild raptor or encourage your moggie, if you have one that doesn't resemble Garfield.

158

Possums

Possums are native animals in Australia and, despite being
numerous and pestilential in some areas, they are protected.
I have harboured a strong dislike for them ever since a brushtail
possum developed a taste for the leaves of my apple tree. I was
fond of that tree before it was systematically stripped bare.

Check with your local council to ensure they permit problem
possums to be trapped and relocated and, if so, buy or rent a
possum trap. Apple makes a good bait. Rub it on the bark of the
tree they inhabit down to the ground near the trap and place the
apple in the trap.

As for New Zealand, possums are an introduced pest. No
constraints for you guys, but best to seek advice on humane
methods of dealing with them.

8

CHICKEN FEED

G one are the days of buying sacks of different grains and mixing up a stockfeed for your hens: animal husbandry has moved on from that a bit. You can just mosey down to your local stockfeed supplier or, in an increasing number of areas, supermarket and buy a premixed food in grain, pellet or mash form. In fact, chicken feed ain't just chicken feed: more food science goes into it than you might think.

Commercial feed

Your modern layer mix has benefited from decades of dietary study undertaken by the commercial egg farming industry, aimed at increasing the egg-yield-to-food-weight ratio. This study of diet has gone hand in hand with commercial breeding and resulted in hens that will reliably lay an egg a day, year round, if their diet is right. The downside of this relationship is that your modern high-performance laying crossbreed hen is a lot more sensitive to diet than its purebred forebears.

Imagine feeding an Olympic athlete on weight-loss cuisine or, at the other end of the scale, double cheeseburgers and fries for every meal. That's the end of personal bests, I'm a-thinkin'. Hens are

Opposite: Chickens love a treat of whole grains but, if you give them too much, egg laying might decline.

the same: to make an egg every day, they need a diet that includes protein, calcium, trace elements and other components. The proportion of these can vary a little but if it varies too far they will stop laying, especially high-performance layers and older hens.

What about the hens you see foraging for themselves in Asia and Africa, I hear you ask. They are surely not fed on commercial stockfeeds? Quite true, but they don't lay an egg a day either. Many purebred hens are quite able to survive on forage, kitchen scraps and the odd handful of grains thrown in their direction, but a performance layer has too high a metabolic rate to survive on forage. She needs high-octane fuel to keep that production line going and, without it, the first thing that happens is she'll stop laying.

I had one customer who decided for some reason to allow her ISA Brown hens to survive on forage only. One starved to death and the other only recovered with careful attention by me. This was in a large garden too, with plentiful and varied plants to eat. I was pretty upset about it, I can tell you.

164

Annoyingly, I still occasionally discover feed marked as 'layer pellets' with as little as 10 per cent protein. People buy crossbred laying hens, feed them with this stuff and then wonder why they are not getting any eggs. It's a trap for younger players. Check the protein content of the feed and if it has less than 14 per cent, find another product if you want regular eggs.

Calcium content (for making eggshells) is also important – it should usually be in the 3–6 per cent range. Feeds at the upper end of this range are better suited to older layers who are less able to metabolise calcium. Although, too high and it can be toxic to them, so there are limits.

Believe it or not there is still more to feeds. Stockfeed manufacturers sell to commercial egg farms with hundreds of thousands of hens either in cages or big barn setups. But they might take a few tons, put it into 20 kg (44 lb) bags and sell it to those mad fools who keep hens at home. But for us fools who keep hens at home, if we buy that feed, it assumes the hens are in a commercial environment and that the feed is all of their diet.

Backyard hens have a much more varied diet than commercial hens. That is one of the reasons their eggs taste so good. For this reason I recommend that you seek out feed formulations that are intended for the 'backyard layer', or similar. The ratio of ingredients will be different and should assume that a not insignificant proportion of the diet comes from their environment, whether or not you feed the hens kitchen scraps. There are enough of us fools around now to cater for and there are various feeds available.

Pellets, crumble, grains or mash?

By eating a pellet, the hen is getting the whole diet, all the components. While feed for egg farms is almost always in pellet form, pellets don't automatically mean feed formulated for farmers. There are now some backyard mixes available as pellets.

Crumble is basically pellets that have been chopped up and made smaller and easier for the hens to eat – it is the most common form of feed for starting and growing chicks. If you decide to feed your hens pellets or, more particularly, change their diet from other feed to pellets, you might find that the hens don't recognise it as food. If you want to turn your pellets into crumble, throw them into the blender – *et voila!*

While on the subject of changing feeds, you don't want to suddenly change the hens' diet or there is likely to be an interruption in the supply of eggs. Changing feed is best done by gradually tapering off the old feed as you blend in the new in increasing proportions.

Personally, I'm not big on pelletised food. It just doesn't have the natural feel about it. Kind of like space food for chooks – nutritious, sort of tasty, but epitomising inoffensive and dull.

Grain feed is the real deal: proper food for hens and they love it. But grain feed has a dark side. How do you add the necessary extra calcium, trace elements and protein to grains? Some feed manufacturers will put all that good stuff in a pellet mixed with the grains. When you include the pellets in the breakdown of the dietary components in the whole feed, the proportions are right. The house of cards is reliant on the hens eating the pellets and, given any choice, they won't. Other manufacturers add the good stuff to a powder that is mixed through the grains and the hens eat the grains and leave the powder – same problem. I have heard of the addition of an oil that helps the powder coat the grains, but I have my suspicions about its efficacy.

For these reasons, I use grains as a supplementary feed, not the core of the diet. See page 172 for information on scatter grains.

165

This just leaves mash. Mash is grains and the good stuff put through a mill. The grains are chopped up and therefore not selectively targeted by the hens. The birds tend to get a beak-full that includes all the elements of the diet. Mash can be dry mash or wet mash, where you add water. Most mash is dry mash, so it is best to check with the supplier or manufacturer before mixing in water to make a wet mash.

You might have guessed that mash is my preference, but now you are able to make your own informed choice.

Organic feeds

Why would you want to pay more for organic feed? Because, to gain certification:

- Farmers must nurture their paddocks in a sustainable way that naturally improves the soil;
- Farmers are required to maintain the local habitat and promote bush regeneration, control weeds etc;
- Every grain used to make up the organic feed must be certified organic;
- The process of mixing and milling the feed must also pass certification; and
- Pesticides, herbicides, growth regulators, hormones, antibiotics and other nasties are totally prohibited.

There are more reasons to go organic but I must curb my enthusiasm here for I fear I may bore you rigid on the subject.

Kitchen scraps

Hens love kitchen scraps; the eggs taste better, and feeding scraps to the chickens dramatically reduces the amount of waste that goes from your place to landfill.

Chickens are opportunistic feeders and omnivorous and within some broad parameters they can eat anything you can eat. But some foods contain substances that are poisonous to hens. Some pulses (beany things), for example, and foods with high salt content can be poisonous to hens but in moderation and mixed with lots of other stuff you can get away with them.

Don't give your hens leftovers from food preparation that contain toxins and are poisonous to humans, such as rhubarb

Opposite: The pullet and the pellets. Many starter and grower feeds are supplied as pellets.

leaves (the stems are ok) and green potato peelings (non-green are ok). I have also read not to give them avocado, chocolate (as if!) and coffee grounds. As for coffee grounds, I like to imagine the hens displaying the behaviours of your over-caffeinated office colleagues – irritable, snappy and needing that pointless report *immediately*!

I have been feeding my hens leftover everything and anything from my kitchen for over 20 years and have never had a problem. This includes meat and residues of dishes that included meat, such as spaghetti bolognese.

But I am quite careful that older food from the 'land-at-the-back-of-the-fridge' doesn't include food from the land that time forgot. Early in my chicken-keeping days I was told by a reputable informant that the really old and mouldy stuff can give the hens a crop infection, so I turf that. Problem free for over twenty years, I had no reason to suspect my practices. Well, not until I came to be writing a book on the subject, when I decided to ground my position and unwittingly opened Pandora's Box.

What is a kitchen scrap, I ask you, and why should it matter? It doesn't so much – unless you happen to be in the UK. Well, more specifically, in England, Scotland and Wales, where there is a prohibition on the addition of animal-based proteins to the diet of farm animals as part of measures to control disease. By the UK government definition, farm animals includes pet poultry, meaning that if you live in the UK it is illegal to feed kitchen scraps to your hens, regardless of whether the scraps are animal or vegetable in origin. This doesn't apply to vegetable matter straight from the garden to the hens: it just can't pass through your kitchen, unless you are a vegan, apparently.

Some argue that this is a ridiculous overreaction and that stockfeed suppliers lobbied the UK government for their own nefarious purposes during the development of these regulations – surely not? I choose not to comment. Although, it should be noted that the UK has been a little over-represented in food-based disease outbreaks over the years, with problems such as mad-cow disease, foot and mouth and salmonella.

In Australia, the Queensland government has banned swill feeding of livestock. This applies to the feeding of kitchen waste that incorporates animal matter (or vegetable waste that has been contaminated with animal matter) to poultry. At the time of writing, it was hard to get information as to whether this applied only to farmed animals, but the legislation is administered by the Department of Agriculture and Fisheries, suggesting farmers as the target.

For the rest of us: meaty feed or no meaty feed – it's up to you.

Garden forage

Being omnivorous, hens will have a go at anything that moves and quite a lot that doesn't. Anything they can scarf down really – small lizards, ground-dwelling spiders, centipedes and other crawlies – all form part of their diet. I have had many customers from tick-infested areas say that since the arrival of the hens the ticks have scarpered.

Hens will also scratch around for seeds, grains and so on and, if penned in an area, will very effectively turn over a vegetable patch at the end of the season or completely denude it of annoying, recurring, soft-leafed weeds.

Bugs and garden greens are an important part of the hens' diet and you will find that they forage for it as naturally as breathing. If your hen run gets denuded of greenery you will find the birds take a great interest in any weeds you chuck over the fence for them, as well as vegie matter from the kitchen.

Equally, if the hens have plenty of green forage, they will take less of an interest in vegie kitchen scraps. This is worth a mention because I have had many people ask me what they are doing wrong when their hens take no interest whatsoever in vegie material from the kitchen. The answer is nothing. Please also remember that these are live animals and they have (small) minds of their own.

169

Overleaf: Garden forage for hens – cheap and plentiful.

Scatter grains

Universally, in my experience, hens love whole grains, seeds, corn, and similar foods. Accordingly, there is a temptation to give them lots of it. Restrain yourself! As mentioned earlier, the more eggs your hens lay, the more sensitive they are to diet. If you treat them with lots of grains, they may not get enough calcium, protein and other nutrients and you might start missing out on eggs.

A handful of grains a day is plenty and if you have a small hen run, a cool way to keep the chickens from getting bored is to bury the grains under mulch and make them look for them.

One final word on scatter grains: it is best not to scatter them on the lawn. That is, assuming you don't want a lawn comprising wheat, rye, millet or whatever the grain mix is made of. They are seeds, after all.

However, if you do allow the seeds to germinate there is a funky way to feed the hens fresh yumminess in an ongoing way. Buy a sheet of welded reinforcing mesh with squares about 2.5 cm (1 inch) – the mesh should be stiff enough for the hens to walk on. Mark out an area of the garden the size of the mesh and plant the seed mix in this area – scattering the seeds and raking them in is fine, assuming the soil is broken up. Lay the mesh over the area, propped up on house bricks or similar to raise it a few centimetres off the ground. Give the area a good watering and wait. Once the seeds germinate and grow, the mesh stops the hens digging out the seedlings but allows them to graze on the plants that are sticking up through it.

Storing feed

First and foremost, feed needs to be stored in a robust container to keep out vermin. Better still, go for a container that is airtight (or as close as possible) so the feed will keep longer. Don't even think of buying feed from a pet store in tiny little 5 kg (10 lb) bags unless you are sending Jeeves to get it in the Bentley. Bags of 20- 25 kg (44–55 lb) from the stockfeed supplier are the way to go. If the hens are not wasting much and you are not feeding the entire neighbourhood's pigeon population, a bag should last around three months for two hens.

A word on reusing containers. Obviously, those that originally contained foodstuffs are the best, but you can use old chemical

172

Opposite, clockwise from top left: A recycled chemical drum makes a vermin-proof food storer; Mash feed; Scratching for forage; Serve scraps on something that allows them to be taken away at nightfall.

drums as well, as long as the chemical wasn't toxic. I have used old dye drums from printers and even drums that once held corrosives. However, I did make sure the corrosive was water soluble, and that the plastic was inert enough to ensure no residue could become embedded and leach out later. I gave them a pretty good wash, too.

A standard 60-litre (13-gallon) plastic or metal old-style rubbish bin from the hardware store will do if a reused drum isn't available or is unpalatable, but I encourage reuse and it has always had a role in parsimonious chicken keeping.

Feed keeps well, especially in airtight containers and when the humidity is low, but must not be allowed to get wet. Wet feed rots, and can make your lovely ladies properly sick. You can give wet feed to the girls immediately if they will have it, but don't keep it. I have noticed with pellets, crumble and mash that the wet bits will clump together a day or so after wetting, leaving unharmed dry feed. The clumps can be removed but it is a false economy to try to save too much. Best to err on the safe side and be really sure you got everything that came into contact with the water by taking the clumps and some of their surrounds. If whole grains get wet, plant them!

How much is enough?

An adult hen will consume approximately 120 g (4 oz) of feed per day, although, actually, the total is totally irrelevant. Much more important is the proportion of different feed components and that the hens can always get food when they want it.

In terms of proportion, I mean the right mix of kitchen scraps, scatter grains and feed. The nutritionally balanced stockfeed (mash, pellets, whatever) is the most important. The hens need to have ready access to this throughout the day, especially in the morning. Hens eat most of their diet between dawn and midday, so this is the time to ensure they can get to the feed.

Scraps and scatter grains are of lesser importance than the stockfeed if your focus is plentiful eggs. And, if ever there is an otherwise inexplicable reduction in laying, I recommend you withhold the scraps and grains for a while to see if one or more of your hens is overly enthusiastic about them.

Can you overfeed a chook? Not really, if the hen is active

and laying regularly. Laying is such a metabolic marathon that a good layer can barely get enough to eat even when food is in abundance. A gun-layer is likely to be scruffy, scrawny and actively looking for food: she has no time for frivolities such as plumage. Your plump, indolent and impeccably turned-out hen, languidly strolling the hen run – she's more likely to be a lazy layer.

There are some foods that will make a hen fat and in colder climates some people will advocate feeding the hens these to help them through winter. The hens don't need this stuff, even in winter. A quality feed, some scratch grains (if you want), forage from the garden, and well managed scraps and you'll be fine.

As a guideline to the minimum, a coffee cup full of feed should see a hen through to tomorrow. A chicken should never be allowed to run out of feed, meaning there is no maximum as long as you don't waste it. There you go: feeding sorted.

HEALTHY HENS, HAPPY HENS

Prevention is better than cure

It is genuinely easier to prevent (or at least minimise the likelihood of) problems with chickens than to rectify them. Many of the diseases and infestations that hens can get are present in their environment all the time. The reason they don't get sick or infested is that they are healthy and strong and can prevent the problem organisms from taking hold.

Chickens are sensitive to stress. Stress is like Kryptonite for Superhens and allows them to become overpowered. Hens become stressed by sudden changes to their environment or diet, so try to make changes gradually. Hens are also stressed by overcrowding, boredom and having to compete for food, so a well-managed flock, coop and hen run will make their lives happier and involve less work for you in the long run.

Here endeth the sermon; you can take off your Sunday best now.

Opposite: A healthy
and very happy
Speckled Sussex.

Healthy management

It is my experience that a little regular effort to prevent the hens' environment becoming manky can lead to years of *not* having to deal with some of the problems in this chapter. No guarantees though: I have had customers who have never had mites in their hens and others who got a bad infestation within weeks. Sometimes it is the luck of the draw.

If you put a little work into the healthy management of your hens, you will enjoy them more, they won't smell, the neighbours won't complain and the chickens are much less likely to suffer from problems that can suck the joy out of keeping them as pets. A clean, dry, well-ventilated coop and clean and abundant food and water is disease prevention in its own right.

Keep an eye on your hens, and not just to keep their plotting and scheming in check. Hens' behaviour is a telegraph for their health and if they start behaving differently it is likely something is amiss. Active, alert, engaged, curious hens are almost certainly in fine fettle; whereas listless, unresponsive, slow-moving or stationary hens are likely to be sick. It is not always as clear as that though and any change in behaviour should elicit investigation or, at the least, increased vigilance. Many diseases will infect the whole flock if left unchecked and knowing your birds will allow you to isolate and treat a bird before she becomes very sick or infects her sisters.

Don't be disheartened by the myriad ailments listed in this chapter (and I have only covered the more common ones). I have included reference to these ailments because the book would be incomplete without it. In truth, hens are remarkably robust critters and not prone to disease and infection if they are not overcrowded or stressed.

Finally, if you keep hens for long enough, you will have a hen that dies for no identifiable reason. Don't wig out. It happens. If your other hens seem healthy enough, put the matter out of your mind. The hen can be buried – have a little ceremony if you like. Alternatively, as a dead hen is not greatly dissimilar to a take-away roast chicken, except that it should never be eaten, its remains can be placed in the household garbage. I wrap them in a few layers of supermarket shopping bags (which seem to be abundant no matter how many times I refuse them and how many uses I find for them).

180

Parasites

Collectively called ectoparasites, there are numerous species of insects – such as mites, lice, scaly leg mites – that bother the outside of hens. Some come from wild birds (often pigeons), some arrive in your flock because you weren't careful enough checking newly purchased hens, and some were there all along. Some are more problematic in cold weather, while others reproduce rapidly when it is warm.

I am not going to go into details of different insects because I'm going to recommend you manage them all the same way. Healthy hens can usually keep these parasites at bay by dust bathing and preening. You can further reduce the likelihood of infestation by sprinkling sulphur powder, powdered diatomaceous earth, or a bit of good old lime in the dust baths to make them inhospitable for bugs. However, if the birds get stressed, or if an infestation takes hold in one bird, it can spread to the whole flock very quickly.

How do you identify an infestation in your flock? In short, you will see them on the birds, on perches or in the nest: little black or white dots which may or may not be moving. Sometimes, you will feel them on your arms after handling the birds or collecting the eggs. You might also see the birds more actively preening or pecking at their feathers. (Feeling itchy yet? I saw you scratching.) The good news is that the sort of bugs that infest poultry don't infest people as a rule, but if you get a really bad infestation, you can be impacted too, so best to nip it in the bud.

There are a number of insecticides suitable for poultry, but fewer that are suitable for laying hens. I will elaborate later on the use of medications, and this applies to insecticides and disinfectants as well.

The art of successfully dealing with an infestation is to do it properly. Do the birds (*all* of them) properly and do their environment properly. If you don't, you will just end up having to do it all again.

The first rule with insecticides is to use as directed and that overrides anything you read here. Pesticides are not chemicals to mess with. Don't be tempted to make a 'strong mix' or to mix different pesticides because you think your infestation is worse than the product was designed for. You could make your hens, or yourself, sick.

181

Philosophically, I gravitate towards the benign end of the spectrum when it comes to chemical use, preferring products that have no withholding period for eggs. For parasite infestations, I use a powder called Pestene, which relies on a plant-derived active ingredient called rotenone. It has no withholding period for eggs. I don't know how universally available Pestene is, but rotenone has been used for this purpose for ages. But any product that specifies it is intended for mites and lice in poultry will do.

When dusting the hens, make sure you get them all: either catch them all, treat them and place the dusted ones back into the coop; or take them from the closed coop one by one in the morning and release the dusted ones. They are easier to handle and manage in the evening when they are naturally sleepy and dopey. Grab your patient by the legs and lay her down in the dusting zone. By holding her by the feet you can roll her over with ease to ensure full coverage. Dust her all over, giving special attention to under the wings and around the vent (technical term for a chicken's bottom). Rub the dust in among the feathers: it requires contact with the insects to work properly. When you let her go, she will most likely fluff the feathers, creating a cloud of dust. You might be forgiven for thinking, effort wasted, but I suspect this helps push the powder through the plumage.

When you have done all the hens, it is time to attack the coop. Firstly, pull out, sweep out, or otherwise remove any loose material and bin it. This includes all nesting material. Don't keep it, even if it was recently renewed, and don't compost it: get it off site. Dust the coop thoroughly, with special attention given to little nooks and crannies a bug might want to live in. If you are worried about the bugs getting on you, applying mosquito repellent beforehand works well for me. Next step is to dust the depressions the hens regularly lay in and dust bathe in. Don't be too stingy with the product: do it once and do it well.

While I've just described dusting, the process applies to sprays as well. You just want to make sure you have treated every space likely to harbour the bugs.

Opposite: An older, but very healthy and alert, ISA Brown. Overleaf: Dust bathing keeps parasites at bay.

Healthy hens, happy hens

Scaly leg mite

Blasted things, scaly leg mites: little mites that get under the scales and lift them up, leaving the birds looking awful. Getting rid of these requires perseverance. You have to make an oil-based concoction and either dunk the hens' legs in it, or paint it on. The oil smothers the mites. Do this repeatedly, at least twice a week over a few weeks to be sure, although daily can't hurt if you can be bothered. The quickest way to go about this is at night when the birds are sleepy and on the perch: pick them off, dunk their legs in the concoction and then put them back where they were.

Really work the concoction in under the scales for best results. The concoction? A mix of any kitchen oil and kerosene (more oil than kero, but the ratios aren't that important), and add some essential oils such as eucalyptus, tea tree or lavender oil, if you like. Alternatively, use melted petroleum jelly (Vaseline) mixed with sulphur powder. Whatever takes your fancy. Just don't use car engine oil: it contains nasty additives.

Dropped scales can reinfect the birds for about a month, so to be sure you get on top of it, clean out the coop weekly and add lime, sulphur or Pestene to make the floor inhospitable. Discard the floor covering.

Common infections

I mentioned previously that there are hundreds of infections and ailments that can affect chickens. I also mentioned that good management will dramatically reduce the likelihood of these problems in your flock. For this reason, rather than cover lots of ailments I have had no experience with, I'm going to stick to the ones I have come across in 15 years. The philosophy being that if I haven't come across it, it can't be that common.

For a lot of the nasty diseases, such as Fowl Pox, Infectious Laryngotracheitis, Marek's Disease and Newcastle Disease, there are vaccinations available. In some jurisdictions these vaccinations are mandated for commercial breeders or are routinely given. Breeders of purebred varieties are less likely to comprehensively vaccinate their birds but, like driving without comprehensive insurance, it doesn't increase your likelihood of having a crash, just the financial consequences if you do. Many people drive their entire lives without having an at-fault accident.

Often you will smell the onset of an infection before you see it: sick hens smell bad. Other signs of infection are swollen eyes, snotty noses, coughing, sneezing and lethargy.

If you suspect illness, see injury or observe the symptoms on a hen, it is time to put the affected bird in the sick bay. Sick and injured hens are often picked on by their peers and, if nothing else, separating the hen will minimise the stress on her. The simplest sick bay is a large cardboard box with mesh or something over the top so she isn't tempted to hop out. Put fresh shavings, newspaper or other bedding in the bottom and give fresh food and water. A stress-free chill-out zone is what you are looking to create.

Coccidiosis

By far and away the most common ailment affecting chickens is Coccidiosis (I call it Coxi). It is caused by protozoa in the hens' gut. Pretty much all hens have it in some form or other and a low-level infection builds their immunity, assuming they don't get a secondary infection while they are getting over it.

It is more likely to become problematic when you add birds from a different source as there are many species of Coccidia and different birds may not have developed resistance to a variety they are newly exposed to. However, if the hens are otherwise healthy and stress free, they will usually build resistance.

Coccidiosis is devastating to chicks, and for this reason many starter feeds include a coccidiostat (Coccidiosis medication). A mild Coxi infection may be completely unnoticeable, with the possible exception of reduced laying. A serious infection can kill a hen within 24 hours of the first sign of symptoms.

In my experience Coxi usually shows up with one bird looking especially sick: the infected hen. A badly infected hen looks like a fellow suffering from a man-cold. She just sits or stands there, lethargic, looking miserable and hoping to milk some sympathy. Fluffed feathers and a droopy pale comb are other common symptoms, along with diarrhoea. Because Coxi oocysts are present in chicken poo, special attention to hygiene after an outbreak reduces the likelihood of other birds getting sick from it.

Now I've given you the heebie jeebies, what's to be done about it? Well, I won't gild the lily: the infected hen has an even chance, at best, of survival. Her chances are improved by quick action.

187

While there are a number of coccidiostats available, I have had comparable results with good old garlic. I can't recommend the restorative powers of garlic enough: it is a panacea for poultry, has no side effects that I've been able to discover, and has no withholding period. Garlic has a number of sulphur-based compounds that do wonders for the hen and make life unpleasant for the protozoa.

Treat the infected hen with fresh garlic, crushed or finely chopped. Don't be tempted to go to the jar in the fridge – it's fresh garlic that has all the good stuff. Take a teaspoon, open her beak and pour some in. It is easier as a two-person job, as you can then hold the head back and pour it down her throat. One crushed clove should do. If she is not drinking, spoon in some water too.

Do this twice a day until she shows signs of recovery and then wind it back to once a day for the next few days. Use her response as a guide. If she shows no signs of recovery after a week, she has probably had it and you can let nature take its course, or she may well recover on her own.

For the other hens in the flock, boost their immunity by peeling a clove of garlic and dropping it in the waterer and any other water sources they have too. The water leaches the goodness from the garlic clove and turns into a garlic tea or tonic. Garlic is so benign, I recommend you put a clove into the water from time to time as a preventative.

There are other coccidiostats available from pet shops and farm suppliers, but beware: some are not suitable for laying hens as they contain residual components that can end up in the eggs. Read the label. What is the point of getting your hen healthy and laying again if you only have to throw away her eggs?

As a rough guide, in my experience, about half of the hens badly infected with Coxi will end up in the great chicken coop in the sky, no matter what you do, and of the ones you treat about half will recover fully. So, it is worth giving it a go.

Chronic Respiratory Disease (CRD) and Infectious Coryza

Both CRD and Infectious Coryza are respiratory diseases caused by bacteria. They are rarely fatal (unless a secondary infection sets in) but result in (sometimes very) sick hens and reduced egg production. You won't be in any doubt if your hen has this: classic cold and flu symptoms including snotty noses

Opposite: Active, alert and curious – a healthy gold-laced Wyandotte.

(the holes either side of the top beak), sneezing, coughing and, in the case of Coryza, swollen face can occur. And a stinky odour to boot.

If it is CRD, and a mild case in otherwise healthy hens, they will recover on their own in six weeks or so. Garlic helps, as do leafy greens such as rocket and parsley. If the hens look really ill, and especially if you have a large or valuable flock, it might be time for the vet, as all the hens are likely to end up with it. The vet can prescribe antibiotics that will aid recovery markedly.

The bad news is that the hens remain carriers of the bacteria even though they have recovered from the infection. In a breeding or commercial environment, this is a big deal. To get rid of it permanently means culling the infected flock, thoroughly disinfecting the coop and leaving the environment poultry-free for a month before restocking. In a backyard setting this is unnecessarily drastic. However, if you have a flock that has had it, you might choose to let your flock die out and then disinfect and let it lie fallow before getting new hens.

Avian Influenza

There are a number of strains of Avian Influenza that are hard on the chickens but not transmissible to humans. Luckily, Avian Influenza outbreaks are not common. However, this book would be incomplete if mention was not made of the H5N1 strain which is potentially lethal to humans.

Many years ago, at the time of the major outbreak of H5N1 in Asia, I lost a bit of business due to fears that people's backyard chickens could potentially infect them with a deadly virus. The outbreak in Asia spread to Europe and Africa and people were properly frightened. I was worried too, in part for me and in part for the business. I had some research to do.

To be infected with Avian Influenza you have to be in direct contact with infected live or dead birds. This kind of direct contact is more the sort you have when you are working with chickens in a commercial farm or living closely with them, as is more common in Asia. It is less the sort of contact that we have with our pet hens (unless you invite them into the house).

In other good news, H5N1 (and a second potentially deadly strain, H7N9, identified in China in 2013) has never been recorded in Australia or New Zealand. We benefit hugely in such matters by being surrounded by lots of water (and having strong quarantine and biosecurity requirements). Additionally, the World Health Organization is all over Avian Influenza and if there is an outbreak, pre-existing plans will be implemented to contain it and local agricultural authorities will also step in.

Finally, agricultural authorities are well aware that there are thousands, perhaps millions, of us 'chooky types' out there in towns and cities. If there is a risk to us, we will be told how to manage it with plenty of warning. Your chickens might be plotting to kill you, but Avian Influenza won't be the way they succeed.

Other health problems

Egg bound

Like trying to deliver a ten-pound baby, it is hard work to push out an egg and sometimes the egg gets stuck. Egg bound is more common in older hens because they are more likely to be laying the big eggs that get stuck. I can just see a few readers wincing at the thought.

You will see it with a hen obviously trying to lay an egg but not getting it out. She has to get it out: if she doesn't, other eggs will bank up behind it and she will die. Warmth can help loosen up muscles. There are lots of ways to apply warmth: you can wrap a half-filled hot water bottle in an old towel and put it under the hen. Alternatively, microwave a slightly damp old towel, hold her over a pot of steaming water, or use a heat lamp if you have one. Food-grade vegetable oil can be used to lubricate her vent.

Gentle massage can also assist and you will likely be able to feel the egg from the outside. Don't press too hard – if you break the egg you run the risk of peritonitis. If the egg does break, try to gently extract the pieces. If she is a favourite it might be worth a trip to the vet.

Crop bound

Sometimes food, especially long skinny food, such as long grass or string (if they are dopey enough to eat it), gets stuck in the hen's crop. The crop is the first stopping point in the digestive system in the lower neck. You will notice the crop distended and it will feel firm to the touch.

A little warm cooking oil or olive oil poured into her throat will help, followed by some gentle kneading to loosen things up. You might have to repeat the process – if you do, wait an hour or so between applications.

Bumblefoot

Bumblefoot is an infected injury in the footpad of the hen, causing pain when she walks, so she will limp. If you see a hen limping, check her footpads. If you see an abscess, give the foot a thorough wash with disinfectant. Lance the abscess and squeeze out the nastiness, then dress it with some antiseptic cream and a bandage (a cloth sticking plaster or two should suffice).

Worms

Use of a wormer for poultry is the only preventative medication I recommend. I worm the hens at the change of the season as a memory aid. If you don't want to do this every three months or so, garlic is a homeopathic alternative. Drop a clove of fresh garlic into the water – or add a little powdered garlic to the feed (although fresh is better). Garlic is benign, so you can give it as often as you like.

There are liquid or tablet-based wormers available. I prefer the tablets for small flocks, as you can be sure each bird has had a proper dose – how useful are mixing directions for 'dosage per 100 birds' for someone who has three hens in their backyard? To give a hen a tablet is easier with two people. Hold her head back and drop it down the throat. Hens don't have much of a tongue, so once it is in, it's in. (Not like a cat, which will wink at you and then spit the tablet back.)

When selecting your wormer, check withholding periods and suitability for laying hens.

Use of medications and pesticides

I'm not keen on use of any medication or pesticide unless absolutely necessary. While some pesticides are relatively benign to humans, they are still pesticides and should only be used to treat infestations. I have read books and articles where people recommend prophylactic use, meaning periodically spraying coops, perches and nests to prevent the rise of infestation. Don't

Opposite: A much-loved Leghorn pullet.

do it: use lime or sulphur powder instead. Never use any pesticide in any way other than described on the packaging. Leaving aside the fact that it is illegal to do so, it is also downright dangerous.

We know a lot more about the effects of chemicals than we used to (I still remember seeing old newsreel footage of a table of diners being sprayed with DDT to prove it was safe – they are all dead now) but we still know less than half as much as we should. Always check that the chemicals you propose to use are safe for poultry and safe for egg-laying hens. Household surface insecticidal sprays are not safe for egg-laying hens and should never be used for managing or preventing infestation. If in doubt, ring the manufacturer and ask; their details will be on the packaging.

Whenever applying a pesticide, err on the side of caution with personal protective equipment. Even if the product doesn't require you to wear impervious gloves, wear them anyway, and always use a dust mask when applying powders such as Pestene or lime – it can't hurt.

As regards medications such as antibiotics, they should be last resort stuff. I recently read about an outbreak of Staphylococcus Aureus (Golden Staph) that was resistant to the antibiotic of last resort. We have no bigger guns in the armoury; the antibiotic A-bomb is now a fizzer. Hens are remarkably good at recovering on their own, and adding a rev-up with garlic and fresh greens will get them to come good more often than not.

Disinfection

Disinfection of a coop is a cast-iron bastard to do properly. Some parasites and bacteria are so hard to get rid of, it is easier to burn the coop to the ground and start from scratch. Luckily, these nasties aren't very common. A thorough wash of the coop and equipment does most of the work and disinfectant just mops up the ones that got away.

If you do need to do a full disinfection, prepare yourself for some hard work. It is easiest to do on a hot dry day so that everything dries more quickly. Firstly, take out all loose equipment such as feeders, waterers and nests, so that these can be washed separately in hot water. Then use a dry brush or vacuum to remove loose material and cobwebs. Wet the coop down and scrub it all over, using hot water and detergent. You really want to remove all traces of dirt, poo and organic matter as these are the places where parasites, protozoa and bacteria live. If you have access to a high-pressure washer, use it, and if you can supply it with hot water, do so.

After washing the feeders and waterers, throw in the scrubbing brush and any other poultry related equipment and give your boots a wash too for good measure. Let everything dry and then use a poultry-suitable disinfectant. Apply as directed. Don't let the hens in until it is all dry and, if you can, house the hens elsewhere for a while as most nasties can't survive long without the bird to live off.

Managing shock and injuries

I have seen hens recover from some horrendous injuries: they are tough little customers. However, the shock associated with the injury is much more likely to have them pushing up the daisies. In fact, a hen is much more likely to die from the experience of being chased and bitten by a dog than by the puncture wounds themselves.

The first 24 hours is critical and the hens need to be taken to the sick bay (see page 187) as quickly as possible: warm, dark, quiet and calm.

You can treat a wound on a hen as you would on a human. The same disinfectants and antiseptics we use work fine on hens. If you have to sew up a wound, you can do it without anaesthesia as chickens are remarkably tolerant to pain. If you don't feel comfortable doing it yourself, your choices are to suck it and see (if she survives the first 24 hours, she will most likely come good on her own) or head off to the vet to have your wallet lightened. If the hen is suffering badly, consider culling.

Extremes of weather

Sudden changes in weather and very hot weather really knocks chickens about. It gives them stress, which, in turn, lowers their immune systems and makes them susceptible to any or all of the above-mentioned ailments. In a commercial environment it is possible to minimise such variables, but the stocking densities are higher and this in turn contributes to stress.

Cold weather doesn't bother hens much unless their coop is not well suited to it. Extended wet weather, especially if it is warm wet weather, is the perfect breeding ground for Coxi. Extra attention to hygiene helps reduce the likelihood of infection when it is wet. Dry, clean floor covering in fixed coops, rotating hen runs (if you can) and moving mobile coops more regularly so the hens are not walking around in their own bogginess and poo pay dividends in hen health.

Thunderstorms and violent weather freak hens out a bit and if you can relocate them to a more sheltered location, do it. If you can't, you can't.

Billy Connolly said: 'There is no such thing as bad weather, there are only the wrong clothes.' Chickens are pretty weather tolerant but they can't take their clothes off when it gets too hot.

Hot weather is the hardest on the hens and I discussed in chapter 5 how to manage that overnight in the coop, but what about daytime?

Hens can't sweat and they regulate their heat by panting and holding their wings away from their bodies. This is normal behaviour and shouldn't worry you, but it is indicative of the hens working to shed excess heat. Hens will start doing this in 30°C (86°F). If you see the hens doing this and you get fewer eggs, or some thin-shelled eggs, in a week or so, that will be the reason.

Your first defence against heat stress is shade: it keeps the ground temperature down. On a 40°C (104°F) stinker, the ground temperature is well over 50°C (120°F) and hens are much closer to that than you are. Tree shade outguns shadecloth and even structural shade from buildings and fences. This is because the trees are lapping up the sunlight to turn it into energy, not just reflecting or absorbing it and heating up. If you have no tree shade in your hen run, consider a temporary hen run under a shady tree.

Water, water and more water: offer your hens plenty to drink, placed in strategic shaded locations. Or you can even freeze an ice-cream container full of water and leave it out in the morning on heatwave days. It will melt during the day and provide cool water in the afternoon when it is most needed. Cool water is more refreshing than warm water and hens' intake of water more than doubles in hot weather.

Over 40°C (104°F) and it is a good idea to be more active in heat management of the chickens. Hose them down or dunk them

Previous page, from left: An Australorp, a silver-laced Wyandotte and a gold-laced Wyandotte looking beautiful. Opposite: It is nothing to worry about if feathers get ragged and scruffy – your gun-layer doesn't have time to fuss with her plumage; she's putting all her effort into laying.

Healthy hens, happy hens

in a bucket of cool water; they won't love you for it at the time, but will be better off for it. If water restrictions permit it, you can also set a sprinkler in the hen run. If you have air conditioning, the hens can be brought inside — a children's playpen or puppy pen and lots of newspaper on the floor will keep them from messing up the carpet. Putting them in the fridge is overdoing it.

Embuggerances

Flies

Overcrowded hens and a failure to clean up poo in the hen run can lead to an outbreak of flies in the run, especially in summer. This can be minimised by using the deep litter mulch method (page 102), but sometimes you get them anyway. Fortunately, they are easy to control. I use an Enviosafe Flytrap. It looks like a plastic jam jar with a funny lid and is basically a lobster pot for flies. They get in but can't get out.

The flies are attracted to a food grade attractant, so no pesticides are used and once a few get in and die the odour from their demise acts like a tractor beam on their peers. It doesn't just clear the garden of flies, it drains the suburb. There are heaps of them in there.

Hang the trap outside because it gets pretty pongy when it is going full steam and when the smell gets overpowering you can chuck the trap into a bucket of water to drown the survivors. Wash it out (stinky task), add new attractant and start again.

Mozzies

I have had customers call me claiming that their hens have fowl pox or the plague or something equally dire that they have found on the Internet, describing spots or scabs on the comb as a symptom. It's just mozzie bites. Some hens will shake mosquitos off their combs and others will just sit there and get bitten. Sometimes the bites will be bad enough to leave black scabs on the combs and wattles.

If you are worried, put some antiseptic cream on the scabs, but if left mozzie-free they will recover. A friend of mine had a show bird with terrible comb scabbing and four weeks later won a prize with her – totally recovered, no scarring, nothing.

The best way to deal with mozzies is to find and empty out any stagnant water, where they breed. For water that you can't drain, and are not intending to drink, a bit of kerosene on top will smother the mosquito larvae. If you feel you have to, you can use citronella candles and diffusers and even good old mozzie coils outside the coop, but be careful the smoke doesn't collect in the coop (and don't burn the place down while you are at it).

Early risers

This is a funny embuggerance, but it happens quite a lot. You have a hen that cackles a bit, early in the morning. So, out of consideration for your neighbours, you leap out of bed, let the chooks out and throw them some feed to shut them up. Over time, you get woken up earlier and earlier until it drives you mad and you call me for advice on how to put an end to your suffering.

You have just been trained by a chicken! She has trained you to get up and feed her at her beck and call. The solution is to get nasty. When she next calls you, don't let her out, don't feed her. Grab the garden hose, give her a spray and go back to bed. I don't mean full blast and knock her off the perch, just a little unpleasant shower. It probably won't work overnight, but give it time. You might get a sense of satisfaction from this, too.

No eggs (plus thin shells and other eggy problems)

There are three reasons for not getting eggs: the eggs are not being laid, are being stolen, or are being laid in a secret location, known only to the hens.

The last of the three is the easiest to solve: go on an egg hunt. A sneaky way to find the nest is to leave the hens locked up for a bit longer than usual so that they are bursting keen to lay. Open the door and watch where they scuttle off to. My record feral nest had 42 eggs in it.

If you suspect your eggs are being stolen, it will be easier to deal with if you can identify your thief. If the burglar is a rodent see chapter 7. Another common thief is a magpie, currawong or crow. These are intelligent, opportunistic wild birds that will steal eggs if they can carry them but will also skewer them and eat them in situ, leaving mess in the nest. This might lead you

to suspect one of your hens of laying thin-shelled eggs or being an egg eater, as I did. The solution was to move the nest boxes from outside in the hen run to deep within the coop. Once they were out of sight they were no longer attacked, identifying the perpetrator and vindicating my hens.

Now we come to not laying. There are a number of circumstances that can result in egg-laying stopping, especially in a backyard setting where the chooks are outside and free to roam. Sudden changes in their environment, such as change in diet, shock, injury or extremes of weather, will often result in cessation of laying.

There is often a delay after the stimuli before laying stops. The chickens lay the eggs they have in the pipe (for want of a better expression) and don't make any more for a while until things have settled down. Hens need to be in good health to lay consistently and not laying is a good early indication of an ailment or parasite.

It is not always possible to identify the reason your hens are not laying but, in truth, it doesn't really matter as long as you can get them laying again. For this reason, I recommend you approach the non-laying as a process of elimination and run through the quiz on page 206.

Thin-shelled eggs from older chickens

A bit more on thin-shelled eggs as a precursor to the end of laying in older hens, who are not able to metabolise calcium as well as they used to. If you are getting thin-shelled eggs, check that the main feed you are giving the birds has sufficient calcium. It should be between 3 and 5 per cent in a laying mix. Making shell grit available to older hens can reduce the incidence of thin-shelled eggs and encourage them to continue laying a bit longer. It should definitely be made available if you are consistently getting thin-shelled eggs but you don't need to rush off to the pet shop – or for the oysters and hammer to make your own – if you've had just one or two. That is more likely to be caused by stress, rather than your ladies having commenced an inexorable decline into barrenness. Give it a while and see if it is a recurring problem. Sudden changes in diet and extremes of weather (especially hot weather) can also result in the hens laying thin-shelled eggs, and thin-shelled eggs can also contribute to egg eating (see page 140).

203

Opposite: Most commercial eggs are from ISA Browns or Hyline Browns, which lay brown eggs. Backyard eggs come in a variety of colours from blue–green to chocolate, depending on your breeds.

Culling

Sooner or later the long-term chicken keeper will arrive at a circumstance where culling is warranted. You might have a flock of older hens that you are feeding but who are no longer laying. Or an illness or injury that requires a hen to be put out of her misery. You might have unwittingly raised a rooster. The reason is irrelevant really: it is your decision.

I often get asked to cull customers' hens and I used to perform the service, but hated it. If you need to cull a hen, take her to the vet and have her put down with an injection. You won't have to be directly involved and it is only money, after all. An honest vet should only charge you the price of an initial consultation if you are clear that you only want the hen culled.

Opposite: The silver- and
gold-laced Wyandottes
find something of interest.

Why no eggs?

| Diet? | → | Check the protein content of the feed (see page 164). If you decide to change the feed to a better or higher protein version, try to blend in the new feed gradually. Reduce the quantity of kitchen scraps you are giving, to ensure the birds are getting enough protein in their diet from the stockfeed. Better still, hold off the scraps completely for a while. |

| Stress factors? | → | Did the neighbours get a new dog? Have you recently added new hens to the flock? If non-laying is related to this sort of stress factor the chooks should return to laying when they are used to the new circumstances. |

| Extremes of weather? | → | This is a common cause of short-duration non-laying as the hens don't eat as much or stop eating completely, especially in very hot weather. You are going to have to wait this one out, I'm afraid. |

| Internal parasites? | → | Use garlic as your first line of attack. Peel a single fresh clove of garlic (don't cheat and use the chopped stuff from a jar) and drop it into the water; repeat for all other water sources in their environment. It is a good idea to worm the chickens now too, to eliminate worms as a contributor. |

| Infidelity? | → | Are your ladies laying their eggs in someone else's nest? Go on an egg hunt. |

External parasites?	→ Check if your hens are infested with mites or lice (see page 181) and treat accordingly.
Respiratory infection?	→ Many of the respiratory infections that affect hens will result in reduced laying. Check for snotty noses. Use garlic in the water for the whole flock and crushed fresh garlic down the throat of any infected birds. This will help with a mild infection of Coccidiosis, too.
Moulting?	→ If your hens have decided to go into moult they will usually stop laying for a few days to a number of weeks. While moulting usually takes place in autumn/winter, it can come at other times and is not always accompanied by noticeable feather loss.
A combination of ailments?	→ The non-laying could be caused by more than one factor, each of which on its own would have had a negligible effect. This is why a holistic approach is a good idea.
Old age?	→ Older hens typically lay fewer but bigger eggs and eventually stop laying altogether. This can be signalled by larger eggs or eggs with thin shells but, equally, it can just happen. She has laid her last egg, that's it. The cessation of laying in older hens can be brought forward by one or more of the above ailments, so it is still worth treating for them. Who knows, she might just start up again once she is well.

207

10

ALL ABOUT EGGS

B ack to the good bits of keeping hens, and eggs are the best bit. Once you have had a proper home-grown egg, nothing else compares. This factor could be contributing to the ever-increasing number of backyard henvangelists around today.

Egg health benefits

The health benefits, or lack thereof, of eggs have long been debated. Eggs naturally contain cholesterol. Ever since studies showed the link between elevated blood cholesterol and increased risk of cardiovascular disease, foods that contain cholesterol have been put in the naughty corner. Unfortunately, as I discovered in school, once you have a reputation for misbehaviour, exoneration is a long and slow process.

The same goes with eggs. It turns out that there is more to cholesterol than you might think. There is good cholesterol and bad cholesterol, and if you don't eat cholesterol your body will make more of it to compensate. You are much more likely to have elevated blood cholesterol from eating saturated fats and trans fats – the sort of fats you get from deep-fried and commercially baked foods. Eggs have very little saturated fats and no trans fats.

Opposite: I collected my first egg from my grandmother's hens at the age of eight and have never forgotten the experience.

Backyard Chickens

- Eggs are proper little powerhouses of nutritional goodness, containing folate, some hard-to-get vitamins and omega 3 fatty acids.
- Extensive studies, involving hundreds of thousands of subjects all around the world, have shown no adverse effect whatsoever on your ticker if you eat six eggs per week. They are not suggesting that more than six eggs per week is bad, just that they can be certain six eggs per week is harmless.
- Cooked eggs are fine for pregnant women and provide heaps of good stuff for growing babies, but home-made mayonnaise and aioli prepared using raw eggs should not be consumed by pregnant woment because of the small risk of Salmonella if the eggs have been poorly stored or handled.
- Avoiding eggs during pregnancy and breastfeeding has no impact on the likelihood of your child having an egg allergy (and the same goes for cow's milk and nuts, for that matter). However, having older siblings or a dog at home reduces the likelihood of a child developing an egg allergy – how interesting is that?
- Eggs make an excellent high-protein substitute for red meat and are included as a component of *all* the recommended healthy balanced diets that I could find. Eggs for any meal, not just breakfast.

Collecting and storing eggs

Eggs should be collected as often as possible; if you see one, grab it. As for transport, I am yet to come across an egg storage container that outdoes the ubiquitous cardboard egg carton. A cunning trick, if you find more than a dozen eggs and you only brought one carton down to the coop, is to turn it over (holding it firmly shut, of course). You can carry another five eggs back to the house in the divots in the bottom.

Speaking of upside down, for longevity, eggs are best stored in the carton with their pointy ends down, leaving the airspace uppermost. Although some people advocate turning the carton occasionally to prevent the yolks settling.

To refrigerate or not to refrigerate? That is the question! We now cross into an area of firmly held opinion, so I will detail the arguments for and against and you can make up your minds.

The for-refrigeration advocates' strongest argument is that salmonella can't multiply in fridge temperatures, so refrigeration is the go. Who wants salmonella poisoning? Not I. There is more to it than that though. In Australia, New Zealand and the US, commercially produced eggs are washed, which removes the antibacterial layer the hen puts on the egg as the last step of laying. Accordingly, it is recommended to refrigerate shop-bought commercially produced eggs. In the UK and much of the EU, however, egg-laying hens are required to be vaccinated against salmonella and the eggs are not washed, thus leaving their antibacterial coating. This is why it is common in Europe for eggs to be stored outside the fridge.

Who cares about commercial eggs? I'm never buying them again now I have my own hens, I hear you say. But the relevance for the backyard chicken-keeper is that you shouldn't wash your eggs if you don't have to. And if you really have to wash the eggs (because one of your birds is roosting and pooing in the nest, or other such problems) they need to be refrigerated afterwards or eaten immediately.

The downside to refrigerating your eggs is that fridges are humid and full of interesting smells and eggshells are porous, having (give or take) 6000 air holes. This means that they can, and will, absorb flavours from the other contents of the fridge. Storing them in the fridge in egg cartons minimises this and is certainly better than the egg tray in the fridge door. Another improvement, if you can be bothered, is to put the carton in an airtight container.

A clear benefit of refrigeration is consistency of temperature. As long as eggs are kept below 20°C (68°F), fluctuations in temperature will have a much greater impact on eggs than the actual temperature they are kept at. So once your eggs have been refrigerated, they should stay refrigerated until use. If you have one of those old-style cool pantries, you have found the ideal egg storage location.

How long do eggs keep? To answer this question I need to differentiate between old eggs and bad eggs. Old eggs are still edible but better suited to baking or dishes such quiches. Bad eggs will make you sick. Fresh eggs are best for poaching and frying because the albumen (egg white) holds together better. With a fresh egg, you will see the albumen hold together and sit firm around the yolk, with only a small portion that is thin and

213

spreading. This can be quite a surprise if you have never had a fresh egg before. As the egg ages, the albumen breaks down and becomes homogeneous. This can also be the case with fresh eggs from older hens.

Another process that occurs as the egg ages is that the membrane separates from the shell. So for hard-boiled eggs, use week-old eggs at least; fresh eggs are a pain to peel.

Ageing eggs also lose some of their liquid content, which evaporates, and the air space increases in size. You can use this process to age test the eggs – handy if you have found a feral nest. Put all the eggs for testing in a suitable, preferably clear-sided, container and cover with water to about twice egg height. The freshest eggs will lie flat on the bottom of the container. Progressively with age, the eggs will tend to rise up on their sides and stand on their pointy ends. When they float it means they are very old and more likely to be bad. You will need to decide your own cut-off point at which eggs should be kept or tossed out.

While I'm on the subject of tossing out, bin any cracked eggs. Or use them immediately, but definitely don't be tempted to store them. Cracked eggs have lost all their barriers to ingress of nastiness. Unless violent gastrointestinal upheaval is your preferred method of weight loss, give cracked eggs a miss.

Bad eggs are very easy to identify: they stink. To be safe from ruining a good dish with a bad egg, I always crack eggs individually into a separate bowl before adding to any mixing bowl or pan, and take a good old sniff.

Dating your egg cartons can help with stock rotation if you get a lot of eggs.

The fridge is not the only potential source of odd flavours in your eggs: this can be environmental too. Unsurprisingly, changes in the chickens' diet will not only affect the colour, but the flavour of their eggs as well. In most cases these changes are gradual and too subtle to be noticed.

However, on three occasions I have had customers describe the eggs from their hens as having an odd metallic flavour to them. The only thing in common with all three properties that I could find was that that they had camellias in the garden. In the case of the last one, the hen run was ringed with camellias, triggering my suspicion and causing me to enquire at the other two houses.

Opposite: The difference between a commercial, bought egg and a really fresh backyard egg when poached – on the left you will see how the albumen of the commercial egg breaks down and spreads.

I'm not saying that the camellias were causal to the metallic flavour. With a sample size of three and a gazillion other potential factors that I might have missed, any statistician worth their salt would lambast me. However, I will say that environmental factors can affect the flavour of your eggs.

On the subject of tainted eggs, despite the fact that the digestive and reproductive systems within a hen are different processes that only come together at the end, there are some chemicals which can end up in the eggs it they are ingested. These include medications.

As regards other chemicals in the hens' environment, the cleaner and greener the better. There are often more benign alternatives to potentially harmful garden chemicals. Pesticides should be minimised and copper chrome arsenate (CCA) treated timber should be avoided. I'm often asked whether lawn chemicals such as weeder-feeders are safe around hens and the answer is, I have no idea. I don't leave it at that though. By law, the manufacturer of any pesticide, herbicide, dangerous goods or hazardous material must put their contact details on the packaging and make available a safety data sheet for the chemical. You can also ask the manufacturer for advice on the safe use of their chemical in a garden containing hens.

Freezing eggs

Don't be tempted to put a whole egg in the freezer unless cleaning is a hobby. If you want to freeze whole eggs, break a meal-size portion of eggs into a bowl and whisk, or don't whisk, I don't care. Before you put the portions into a zip-lock bag or your preferred containment stratagem, you might want to add a pinch of sugar if you intend to use the eggs for baking, or a pinch of salt for other uses. The sugar or salt will help prevent the yolk from thickening or gelling.

Ice cube trays can be used for freezing eggs and they also work well for separated whites and yolks. To re-create a whole (Franken)egg, take two cubes of white and one of yolk. Don't forget the salt or sugar for the yolks before freezing.

Selling eggs

I have never sold an egg in my life. Since starting with my first flock I have felt eggs were a by-product of my hobby and mine to give or barter, but not to sell to generate filthy lucre. Don't worry, you don't have to drink my Kool-Aid. It is legal to sell your eggs. Most jurisdictions have requirements that increase in complexity with the number of eggs you are selling. The imperative being to ensure you minimise your customers' likelihood of projectile vomiting. You can't sell eggs in reused cartons either. Fresh eggs, clean eggs, no cracked eggs and you should be ok.

A handy little trick

Many thanks to Anne Perdeaux for this groovy idea: use the egg shells for starting seeds and growing seedlings. Poke holes in the bottoms to let out excess water, fill with soil or potting mix, add your seeds and then line them up in a carton until the seeds germinate. Plant them out, egg shells and all, but before you do, give a little squeeze so the seedlings can break through the shells and don't become pot-bound. I've tried it and it's a cracker (sorry). The seedlings grow well, probably because the planting out in the egg involves very little root disturbance.

And what if you run out of eggs?

It is a diabolical situation for a long-time hen-keeper to have to buy eggs at the supermarket, but sometimes you might have to. How do you choose? What the feather does *Field Fresh* mean when applied to cage eggs? Whose definition of *Free Range* is legitimate? What does *Open Range* mean? Aren't all eggs laid *Nature's Way*? How was it established that these hens are *Happy Hens* – was the hen polling statistically valid and peer reviewed?

By now you have realised I find the deliberate obfuscation in the classifying of eggs very frustrating. I have serious concerns with a definition of 'free range' that allows 10,000 hens per hectare – one bird per square metre. If you want to be sure your eggs come from hens who live in appropriate conditions, buy certified organic – to gain and maintain their accreditation these farms must meet standards and are audited to ensure they do. Look for the certification logo on the carton.

217

11

CHICKEN BREEDS

Purebreed vs crossbreed

Y ou'll remember I recommended early on that chicken-keeping newbies go for a crossbreed or hybrid layer? But that recommendation only applies to first-timers. Once you have some experience, it's time to spread your wings. (Another appalling chook pun – what's the tally now?)

Generalising terribly, crossbreed and hybrid layers are a lot functional and a little bit beautiful. They are the best layers because they are bred exclusively for that purpose. Downsides include the need to be fed on high-protein diets and a tendency toward a more concentrated (shorter) laying span.

Purebreed hens are often very beautiful and vary in functionality. The wonderful thing about purebreeds is there are lots of characteristics you might want to breed into a hen. Remember that chickens descended from wild little jungle fowl that might lay a couple of eggs a month. From that, with 2000 years to play with, imagine what you could do. If it were me, Franken-chicken here we come: it'd be fierce, weigh 10 kg (20 lb), and it would have four drumsticks, and teeth. We'd have to work to get our chicken dinner.

Opposite: A glossy Australorp scratching for food. Hens will walk forwards, scratch, step back, check, scratch some more – sometimes until they've dug a hole.

Sadly, others are more conservative. But you can still get a hen (Araucana) that lays blue-green eggs that go beautifully with ham. Another favourite of mine is the Plymouth Rock with its lovely banded feathers.

The breeds we have today reflect the needs of breeders past. Some wanted dual-purpose birds that were good on the table but still laid eggs. Hens in cold, harsh environments were bred with rose combs because it is hard to get frostbite in the short comb. Inexplicably, the Australorp, the hen bred for Australian conditions, has jet black plumage which you'd think would be poorly suited to hot weather.

Some desirable characteristics bred into hens include longevity, adaptability, the ability to survive on a poor diet, extended egg laying and disease resistance. Everything is a trade-off though and a strength in one characteristic, such as table weight for meat birds, usually corresponds with poor laying.

There are too many characteristics bred into hens to discuss here. My advice is to find a breed you like the look or sound of, and adopt it – the variety adds colour to your garden.

222

Bantam vs full-size

The diminutive stature of bantams doesn't mean they are particularly better suited to smaller gardens or are gentler on gardens. This varies from breed to breed and bird to bird. Breeders can elaborate for you.

Bantams are a mini-me version of the same breed of hen, although I'm not aware of a full-size version of some bantam breeds. In my experience they are fleet footed and none too keen when it comes to handling by owners, especially children. However, they often make good broody mothers (and correspondingly poor layers).

Being small, they are better at shedding excess heat than their standard-sized brethren and so quite well suited to hotter climes. You might think that because they are smaller they would be picked on by larger birds. This is sometimes the case, but usually sorts itself out once the pecking order is established. They are faster, too, so even if they are bullied, they can get away quickly.

In my opinion, every well-rounded flock should have at least one bantam in it, just to mix things up.

Opposite: This black-speckled Barbu d'Uccle is a Belgian bantam breed.

How to source specific breeds

Breeders of purebred hens don't tend to have websites, although some will share their contact details in online directories. If you are looking for information on specific breeds, I have found it to be readily available in books (plural) on the subject and from breed-specific poultry clubs. Your local library will have books that include descriptions of the features of different breeds.

There is a reward-for-effort component to sourcing a specific breed. The greater the effort, the more you enjoy the bird when you get it home. I take great pleasure in attending major agricultural shows such as the Sydney Royal Easter Show. Wander down the aisles of show birds and find one you like the look of.

Often you will find a breeder, judge or official who (if you catch them in a quiet moment) will very generously share their knowledge with an enthusiastic audience. (You might need an exit strategy though.)

Once you have settled on a breed, you just need to find a breeder. The big shows have a directory of exhibitors and you can buy an inexpensive copy. Don't be surprised if you find the only contact for the breeders of your chicken of choice is a postal address. This is because many breeders of heritage breeds are elderly country folk.

You remember snail mail? I know it reeks of nostalgia, but you're going to have to go and buy a postage stamp. If you do get a yes to your request to purchase one of their hens, it might also entail a weekend away because the breeders are rarely located conveniently on your doorstep.

Some of the more popular breeds, such as Wyandottes, Old English Games and Langshans, have associations that are more contactable and have good member directories of breeders. You could even enjoy the benefits of membership of a poultry club which exceed the nominal member fees they tend to charge.

Some breeds to consider

Araucana

Araucanas originated in South America, but have lost their distinctive features such as ear tufts and floppy pea combs through interbreeding.

They are moderate layers (180–200 eggs per year) and known as the 'Easter egg chickens' because they can produce eggs of different colours, including shades of blue and green. They are very cute too, being small hens with pretty grey plumage.

Araucanas have quirky personalities, don't wander too far and are good foragers. Definitely worth having in your flock for the novelty value of the coloured eggs.

Australorp

The Aussie hen, bred from the English Orpington and suited to Australian conditions. Australorps have black feathers with a green sheen and tend to mature quickly, with some hens producing eggs from five months of age.

They have a quiet temperament, are prone to broodiness and are big eaters.

Renowned for being prolific layers (300 light-brown eggs per year), Australorps have held many world records for laying, including one long-held record of 364 eggs in a year. Australorps are also a meat bird; can this hen do no wrong?

Every flock should have at least one of these lovely birds.

Leghorn

Leghorns originated in Italy and have been immortalised by the Warner Brothers character Foghorn Leghorn. They are very popular, especially the white Leghorn, perhaps because of the large number of white eggs they lay – 300 per year – and because they rarely suffer from broodiness.

Leghorns are alert and active, but not ideal pets as they don't like being handled.

New Hampshire

Originating in the New England area of the US and often called
New Hampshire Reds, due to the rich chestnut-red colour of their
plumage, purebred New Hampshire hens are prone to broodiness
and are mediocre layers (120–150 brown eggs per year). However,
they are often crossbred and, despite looking and behaving like
the purebred versions, the crossbreeds tend to lay a lot better.

These lovely hens make excellent pets as they are very easy to
handle. A good all-purpose fowl: no fuss and fairly placid.

Rhode Island Red

The 'Rhodie' originated in Rhode Island, USA, and is red (blow me
away with a feather). This is one of most popular breeds, bred both
for meat and eggs. The roosters are large and magnificent, but can
be aggressive until you turn them into delicious coq au vin.

These are good layers, producing 200–250 large brown eggs
per year, despite being prone to broodiness and making good
broody mothers. They are greedy blighters and grow quickly into
large hens.

A great garden hen, a good pet, adaptable to a large range of
environments and another must-have-at-least-once breed.

Silkie

The Silkie originated in Asia, probably China or Japan. They've
been around for quite a while: some say that when Marco Polo
returned from Asia in the 13th century he told tales of seeing
Silkies to his mates in the pub. Pull the other leg, Marco – by the
way, your shout!

Silkies are rubbish layers, but top-notch broody mums, with a
strong mothering instinct. Often called bantams, they are actually
somewhere between the two, being smaller than a standard but
larger than a bantam.

One downside to their super-cute soft, fluffy plumage is that
their feathers are not waterproof. So Silkies need to be be kept
under cover in heavy rain.

They have a gentle nature, are calm and make a good pet
for the kiddies. They are also long lived – to 12 years – but can be
fussy eaters.

Opposite, purebreeds,
clockwise from
top left: Araucana;
Black Leghorn; Australorp;
New Hampshire.

Sussex

The Sussex originated in (you'll never guess) Sussex, England. These pretty hens have a collar of black feathers and come in several colour varieties: Brown, Buff, Silver, Speckled and White.

A good layer – 260–280 light-brown eggs per year – not surprisingly, the Sussex is keen on her food.

The Sussex is easily managed and is adaptable to various environments, including small spaces. Alert but not shy and a good free-range forager, this is an excellent backyard hen.

Wyandotte

Oh my giddy aunt – Wyandotte owners and breeders are passionate! They just love them. I have had a few and, truth be told, they are lovely docile hens. Originating from New York State, USA, they come in heaps of varieties, including: black, buff, Columbian, gold-laced, silver-laced, silver-pencilled and white. The beautiful silver-laced are my particular favourites.

Another utility breed, these are used for both eggs and meat. They are reasonable layers: 180–200 eggs per year, although some varieties can be poorer.

ISA Brown

Take a step back, purists! I am aware that the ISA Brown is not a breed, but it might as well be: it's patented. These are not blow-ins; they have been around since the 70s. Not a local, maybe, but worthy of a smile at the supermarket?

ISA is short for Institute de Selection Animale – basically a genetics company. They bred the ISA Brown for commercial egg farms. Said to have been a cross between a Rhode Island Red and a Rhode Island White, it is more likely to be a witch's brew of many breeds these days.

Undoubtedly, the ISA is a gun-layer – 300+ eggs a year – but she also has a reputation for being boom-or-bust in the egg stakes. I have sold ISAs for many years, thousands of hens, and I often get asked how long they will lay for. And I often reply: how long is a piece of string?

As a purveyor of fine poultry, I have a favoured expression: these are animals, not consumer electronics, and nothing is guaranteed. Some ISAs will lay an egg a day for two years, go into moult and never lay again. Equally, I have had ISAs that still lay an

Opposite, clockwise from top left: White Sussex; Hyline Brown; Silkie; Wyandotte.

Backyard Chickens

egg a day at five years old, and steadily keep going until they die –
of exhaustion?

I'm fond of ISAs because they are a great beginner's chicken.
There are a number of similar crossbreeds and hybrids but ISAs
are sold around the world and dominate the commercial egg
market. They are placid, don't get flustered when harassed by
children or pets, and still lay an egg the next morning.

The Rentachook business came about from a henvangelical,
spur of the moment offer to loan my hens to participants in an
eco-living group. 'Try my hens and, when you find that you love
them, get some yourself.' The focus was always to encourage
people to give chicken-keeping a go and, if it didn't work out, they
could give the chooks back. To make this work, I needed an easy-
going, readily available hen that was good for beginners.

Purebreed hens are great fun and I love the passion that
emanates from their owners, but there will always be a place in my
flock for the common-or-garden crossbreed ISA Brown, too.

Opposite: There are
many breeds of chicken;
having a mix in the flock
will make your garden
colourful. Overleaf: The
silver-laced Wyandotte
is as pretty as they come
and a particular favourite
of mine.

INDEX

Page numbers in *italics* refer to photographs.

235

236

237

ACKNOWLEDGEMENTS

There are many people whose help has made this book possible. Firstly, I'd like to thank Pete, my former business partner, for helping start Rentachook. Also, Karl, my current business partner, for running it while I wrote the book and Mel, our trusty chook-breeder, for her beautiful hens. I'd also like to thank my Dad (Tim), the McMurtry and Adams boys, Tom, Ben, Matty, and all the others who have worked for Rentachook over the years.

Thanks, also, to the team at Murdoch Books and those who lent their gardens, children and hens for photography. And finally to my friends and family (Mum and Dad especially), who tolerate my eccentricities, moderate my excessive enthusiasm and keep me sane.

239

Published in 2017 by Murdoch Books, an imprint of Allen & Unwin
Reprinted 2018

Murdoch Books Australia
83 Alexander Street,
Crows Nest NSW 2065
Phone: +61 (0)2 8425 0100
murdochbooks.com.au
info@murdochbooks.com.au

Murdoch Books UK
Ormond House, 26–27 Boswell Street,
London WC1N 3JZ
Phone: +44 (0) 20 8785 5995
murdochbooks.co.uk
info@murdochbooks.co.uk

For corporate orders and custom publishing contact our
business development team at salesenquiries@murdochbooks.com.au

Colour reproduction by Splitting Image Colour Studio Pty Ltd, Clayton, Victoria
Printed by Hang Tai Printing Company, China

Publisher: Jane Morrow
Editor: Jane Price
Design Manager: Madeleine Kane
Photographer: Cath Muscat
Designer: Astred Hicks
Production Manager: Rachel Walsh

ISBN 978 1 74336 753 7 Australia
ISBN 978 1 74336 755 1 UK

A cataloguing-in-publication entry is available from the catalogue of the National Library of
Australia at nla.gov.au. A catalogue record for this book is available from the British Library.

rentachook.com.au